Routledge Re

Economics and Episodic Disease

Originally published in 1991, this study uses the 1983 outbreak of Giardiasis in Luzerne County, Pennsylvania as a case study to explore the social costs of waterborne illnesses to a community. With over 6,000 people affected in that particular case, *Economics and Episodic Disease* emphasises the importance of federal and state drinking water standards to protect the population from contamination whilst also commenting how regulations can be applied to other areas within public health as well as how to appraise the damage caused to surface water by the release of hazardous substances. This title will be of interest to students of Environmental Studies.

Economics and Episodic Disease

The Benefits of Preventing a Giardiasis Outbreak

Winston Harrington, Alan J. Krupnick
and Walter O. Spofford, Jr.

RFF PRESS
RESOURCES FOR THE FUTURE

First published in 1991
by Resources for the Future, Inc.

This edition first published in 2016 by Routledge
2 Park Square, Milton Park, Abingdon, Oxon, OX14 4RN
and by Routledge
711 Third Avenue, New York, NY 10017

Routledge is an imprint of the Taylor & Francis Group, an informa business

© 1991 Resources for the Future, Inc.

Publisher's Note
The publisher has gone to great lengths to ensure the quality of this reprint but points out that some imperfections in the original copies may be apparent.

Disclaimer
The publisher has made every effort to trace copyright holders and welcomes correspondence from those they have been unable to contact.

A Library of Congress record exists under LC control number: 91002129

ISBN 13: 978-1-138-95597-4 (hbk)
ISBN 13: 978-1-315-66590-0 (ebk)
ISBN 13: 978-1-138-95605-6 (pbk)

ECONOMICS AND EPISODIC DISEASE

The Benefits of Preventing
a Giardiasis Outbreak

ECONOMICS AND EPISODIC DISEASE

The Benefits of Preventing a Giardiasis Outbreak

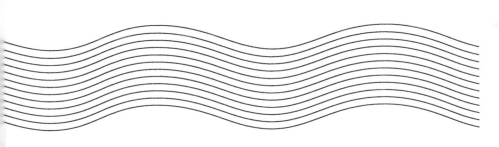

WINSTON HARRINGTON

ALAN J. KRUPNICK

WALTER O. SPOFFORD, JR.

Resources for the Future
Washington, DC

Printed in the United States of America

Published by Resources for the Future
1616 P Street, N.W., Washington, D.C. 20036
Books from Resources for the Future are distributed worldwide by
The Johns Hopkins University Press

Library of Congress Cataloging-in-Publication data will be found at the end of this book.

Parts of chapters 3, 6, and 9 of the present work appeared in a slightly different version in Winston Harrington, Alan J. Krupnick, and Walter O. Spofford, Jr., "The Economic Losses of a Waterborne Disease Outbreak," *Journal of Urban Economics*, vol. 25, no. 1, pp. 116–137. © 1989 by Academic Press, Inc.

This book is the product of RFF's Quality of the Environment Division, Raymond J. Kopp, director. It was edited by Martha V. Gottron and designed by Brigitte Coulton. The index was prepared by Florence Robinson. The cover was designed by Gehle Design Associates.

∞The paper in this book meets the guidelines for permanence and durability of the Committee on Production Guidelines for Book Longevity of the Council on Library Resources.

RESOURCES FOR THE FUTURE (RFF) is an independent nonprofit organization that advances research and public education in the development, conservation, and use of natural resources and in the quality of the environment. Established in 1952 with the cooperation of the Ford Foundation, it is supported by an endowment and by grants from foundations, government agencies, corporations, and individuals. Grants are accepted on the condition that RFF is solely responsible for the conduct of its research and the dissemination of its work to the public. The organization does not perform proprietary research.

RFF research is primarily social scientific, especially economic. It is concerned with the relationship of people to the natural environmental resources of land, water, and air; with the products and services derived from these basic resources; and with the effects of production and consumption on environmental quality and on human health and well-being. Grouped into four units—the Energy and Natural Resources Division, the Quality of the Environment Division, the National Center for Food and Agricultural Policy, and the Center for Risk Management—staff members pursue a wide variety of interests, including forest economics, natural gas policy, multiple use of public lands, mineral economics, air and water pollution, energy and national security, hazardous wastes, the economics of outer space, climate resources, and quantitative risk assessment. Resident staff members conduct most of the organization's work; a few others carry out research elsewhere under grants from RFF.

Resources for the Future takes responsibility for the selection of subjects for study and for the appointment of fellows, as well as for their freedom of inquiry. The views of RFF staff members and the interpretation and conclusions of RFF publications should not be attributed to Resources for the Future, its directors, or its officers. As an organization, RFF does not take positions on laws, policies, or events, nor does it lobby.

Contents

Part 3   Research Needs

List of Figures and Tables

Foreword

Benefit–cost analysis of public projects may be viewed as a decision-making criterion—that is, the project is accepted if the benefits exceed the costs—or, alternatively, as informational input to the decisionmaking process, in which case the decisionmaker employs a criterion more complex than the simple unweighted measurement of benefits and cost. Regardless of the view one has of the use of benefit–cost analysis, most would agree that the technique is growing in popularity, and that it is being applied more intensively than ever and on an ever-broadening set of public policy issues.

For those of us who believe benefit–cost analysis contributes to informed decisions, the growth in its application is comforting. However, the level of comfort is diminished by the knowledge that the quality of the analysis bears directly on its results, and that poor analyses not only fail to improve decisionmaking but may indeed degrade the decisions that they are meant to inform. Unfortunately, poor-quality benefit–cost studies abound; not only do they lead to poor decisions, but they cast doubt on the usefulness of the entire enterprise. The quality of such studies can be raised, however, by educating both their "consumers" and "producers" about the attributes, protocols, and techniques that ensure quality in benefit–cost analysis. In that vein, this book teaches by example.

The book describes the results of an archetypical study from Resources for the Future—that is, a study in which no stone is left unturned and where extreme care and attention to detail have been exercised. Moreover, in traditional fashion for RFF books, its thorough treatment of the subject suggests that RFF has been in the business of examining the costs of public health episodes for years.

The authors have employed a very appealing blend of economic theory and innovative empirical analysis to estimate the social costs incurred by a community as a result of an outbreak of waterborne disease. The results of their research have an obvious bearing on the benefits to be derived from federal or state drinking water standards and the benefit–cost analysis of public water projects designed to protect drinking water from contamination. Beyond that, it is also clear that the economic theory and empirical techniques developed and employed in the study may be usefully applied to other areas, such as food safety, where one might wish to consider the benefits of proposed regulatory programs or to analyze the costs of public health episodes such as salmonellosis outbreaks.

The book also provides information useful for the analysis of costs associated with surface-water or ground water contamination that may come about from the accidental release of hazardous substances. One can easily imagine the theory and procedures described here being applied in Superfund cases where public trustees must quantify the damage to natural resources under provisions of the Comprehensive Environmental Response, Compensation, and Liability Act.

In addition to the specific contributions noted above, the volume reinforces the long-held RFF view that economics teamed with an appropriate dose of natural science provides a very useful framework for the analysis of public policy issues concerning nonmarketed goods and services. It represents a logical extension of our work in benefits analysis and growing familiarity and confidence with the use of survey instruments.

Raymond J. Kopp
Senior Fellow and Director
Quality of the Environment Division
Resources for the Future

February 1991

Preface

Under the Safe Drinking Water Act of 1974, the U.S. Environmental Protection Agency (EPA) is required to establish regulations to protect the public from adverse health effects related to drinking water. By 1983 there was general agreement that *Giardia lamblia*, a protozoan found in some public water supplies, posed such a threat. Although no federal regulation for controlling the incidence of waterborne giardiasis had yet been established, it was under consideration at EPA at the time.

Before the agency proceeded with such a regulation, EPA's Office of Drinking Water wanted more information on the costs and benefits of mandating specific water treatment technologies for water supply systems of particular sizes and with particular risks of contamination. In support of this need, Resources for the Future (RFF) undertook a study, as part of a Cooperative Agreement with EPA's Office of Policy Analysis, to develop methods for estimating the benefits of avoiding an outbreak of waterborne giardiasis, and applied those methods in estimating the benefits of avoiding the 1983–1984 outbreak of giardiasis in Luzerne County, Pennsylvania. The results of this study were used by EPA in 1987, together with analyses that the agency itself had made, in developing the proposed rule for *Giardia lamblia* as part of the National Primary Drinking Water Regulations. The final rule for *Giardia lamblia* was adopted by EPA in June 1989.

This book concerns the measurement of economic losses that an outbreak of giardiasis imposes on individuals, businesses, and communities. These losses include the value of work time and leisure time lost as a result of illness; the costs of medical care; the costs of actions—for example, buying bottled water and boiling tap water—taken by individuals, businesses, and communities to avoid drinking contaminated water; the costs of epidemiological and water supply system surveys; and the costs of temporary measures taken by the water utility to ensure that drinking water supplies are safe. For reasons explained at length in the text, the important losses associated with pain and suffering, with anxiety about the possibility of contracting giardiasis, and with diminished intrinsic value resulting from the loss of a "pure" water supply for drinking are not addressed in this book. And, since our principal focus was the total economic losses that result from an outbreak, to whomever they accrue, three other important areas of investigation fell beyond the scope of this study and thus are not included: namely, an analysis of the incidence of the losses; public policy issues concerning the distribution of costs, equity, and the rights of the public to a safe water supply; and the issue of liability for the losses resulting from an outbreak.

At the time this study was conducted, little was known about the economic losses associated with an outbreak of giardiasis, about the actions taken by individuals, businesses, and communities to avoid drinking contaminated water, or about the impacts on work productivity and leisure time activities of those who had contracted the disease. A case study was needed to learn more about the nature of giardiasis and its transmission and to test newly developed methods, questionnaires, and other survey instruments under real-world conditions. Thus, the case study of the 1983–1984 outbreak of giardiasis in Luzerne County was an important part of this work, and considerable care went into selecting it.

Before choosing Luzerne County, we considered a number of outbreaks in Colorado, Nevada, Pennsylvania, Utah, and Washington. The sites of two outbreaks then going on in Colorado—one at Bailey and the other at Breckenridge—were visited and discussed with epidemiologists and engineers at the Colorado Department of Health in Denver and with physicians, public health nurses, pharmacists, a water treatment plant operator, and residents of one of the affected communities. We also visited the site of the Luzerne County outbreak near Wilkes-Barre, Pennsylvania, and discussed it with personnel at the Pennsylvania Departments of Health and of Environmental Resources in Harrisburg and with others in the affected area knowledgeable about the outbreak.

The Luzerne County outbreak was selected for three principal reasons. First, and perhaps most important, it was large, involving 370 confirmed cases of giardiasis and an exposed population of 75,000. Second, it caused a wide variety of economic losses, affecting individuals, households, businesses, health care facilities, schools, and local governments. Third, it had happened recently, and, therefore, was fresh in the minds of those affected. Indeed, an advisory to boil water was still in effect when the case study began.

This book is intended both for economists interested in the development of methods for measuring the economic losses resulting from nonfatal diseases and for economists and noneconomists concerned with the design of drinking water regulations for waterborne diseases. Two chapters on economic theory are included to provide the theoretical underpinnings for the case study application described in the subsequent four chapters. To make it easier for the noneconomist to follow the development of theory and methods, a special effort has been made in both theory chapters to explain the basis for estimating economic losses and to minimize the use of specialized terms. The remaining chapters in the book concern the nature of giardiasis and its transmission, the case study of the outbreak in Luzerne County, and the implications for policy. These chapters do not require a background in economics, although some familiarity with benefit–cost analysis might be helpful.

This book also may find use as a supplementary text for courses in environmental economics, as it provides an example of how economics can be used to inform the policymaking process with respect to protecting the public from environmental contaminants. The book begins with a public policy issue, and asks what kinds of economic information and data are needed to assist in decisions on that issue. It then proceeds to develop the methods and theory that will be required to provide the needed information and data, and tests those methods and theory in a case study application. The results of the case study lead naturally into a benefit–cost analysis of alternatives to protect the public from environmental contaminants, and lead from the results of this analysis to policy implications, although the policy implications are limited by the limitations of the study. The value of the book in a classroom setting is in the consistency of the example used throughout, in the direct connection between the chapters on theory and those on the case study, and in the comprehensive nature of the example from the development of methods and theory, through empirical application, to guidance for policy.

This study was conducted as a team effort by the three authors. Each author contributed in a significant, but different, way to the study, and without all three contributions the study would not have been as successful. Because of this, the conventional alphabetical listing of joint authors seemed appropriate.

Walter O. Spofford, Jr.
Senior Fellow and
    Project Director
Resources for the Future
February 1991   Washington, D.C.

Acknowledgments

A number of people contributed in important ways to the success of this study. Foremost were staff of the U.S. Environmental Protection Agency (EPA) in Washington, D.C. Alan Carlin at the Office of Policy Analysis and David W. Schnare at the Office of Drinking Water conceived the project and persuaded personnel in both offices of the need for benefit estimates in the preparation of drinking water regulations for giardiasis and other waterborne diseases. Alan Carlin provided support for the project through an EPA Cooperative Agreement with RFF. David Schnare was a consistent source of encouragement and advice throughout the project. We also are grateful to James Hibbs and George Parsons, our project officers at EPA, for their support and encouragement.

Before we could delineate the economic effects of an outbreak of giardiasis and organize the research, we needed to learn more about the life cycle of *Giardia lamblia*, its transmission in water supply systems, and the effectiveness of alternative treatment technologies. We also needed to learn more about the characteristics and symptoms of the disease and its treatment. We were assisted in this early effort by a number of people, including: Dr. Martin S. Wolfe, Office of Medical Services, Department of State; Dr. Charles P. Hibler, director, Wild Animal Disease Center, Colorado State University, Fort Collins; Dr. Ernest A. Meyer, professor of microbiology, and Dr. Edward L. Jarroll, Jr., fellow in microbiology, University of Oregon Health Sciences Center; Dr. Paul Berger, Office of Drinking Water, EPA, Washington,

D.C.; Dale A. Carlson, professor emeritus, Department of Civil Engineering, University of Washington, Seattle; Peter G. Meier, associate professor, School of Public Health, University of Michigan, Ann Arbor; Gary S. Logsdon, EPA Municipal Environmental Research Laboratory, Cincinnati; David W. Hendricks, professor, Department of Civil Engineering, Colorado State University, Fort Collins; Michael Gearheard, U.S. EPA, Oregon Operations Office, Portland; and Mark F. Knudson, City of Portland (Oregon) Bureau of Water Works.

Epidemiologists at the Centers for Disease Control in Atlanta, the EPA Health Effects Research Laboratory in Cincinnati, and state health departments in affected states were a valuable source of information on the epidemiology of an outbreak of giardiasis as well as on the myriad of non-health-related effects. We are especially grateful for the help received from Dr. Dennis Juranek, director, Parasitic Diseases Division, and epidemiologists Dr. Thomas Navin and Dr. Scott Holmberg, Centers for Disease Control, Atlanta; Edwin C. Lippy, EPA Health Effects Research Laboratory, Cincinnati; Dr. Floyd Frost, Department of Social and Health Services, Olympia, Washington; Anne Jensen, Chelan-Douglas Health District, Washington; Jon Baker, Department of Health, Montana; Sandra Capbain, Public Health Department, New Mexico; and Craig Nichols, Utah Department of Health.

Staff of the Colorado Department of Health in Denver deserve special acknowledgment and thanks. Dr. Richard S. Hopkins, chief, Communicable Disease Control, staff epidemiologists Pamela Shillam and Barry Gasbard, and district engineers Joe Meek and Steve Snider all shared their experiences, expertise, and data freely with us during our site visit to Colorado. We are particularly grateful to Pamela Shillam, who not only provided data on outbreaks in Colorado, arranged meetings for us while we were in Colorado, and gave freely of her time, but also helped us conduct a pilot survey of the confirmed cases of giardiasis in Durango, Colorado.

Others in Colorado provided insights into the anatomy of the outbreaks in Bailey and Breckenridge, and we wish to thank them, too. Dr. Thomas Syzek, Crow Hill Family Medical Center in Bailey; Shirley Kelly, public health nurse, Summit County Nursing Service in Frisco; and Dr. Bo Bogan, Frisco Pharmacy in Frisco, all spent considerable time with us and provided valuable information on the outbreaks, the nature of the disease, and the treatment of the disease. Shirley Kelly deserves special recognition for her help in conducting a pilot survey of residents of the affected area in Breckenridge.

The case study of the outbreak of giardiasis in Luzerne County required the cooperation and help of the residents, businesses, municipal and county governments, and the water supply utility in the affected area. Staff at the Pennsylvania Department of Health in Harrisburg were particularly helpful, including Dr. Charles Hays, director, Bureau of Epidemiology and Disease Prevention; Dr. Ernest Witte, director, Division of Epidemiology; and epidemiologists Thomas M. DeMelfi and Bernard Healy. They provided information on the confirmed cases and random telephone surveys of the Luzerne County outbreak made by the Department of Health during the outbreak and helped conduct the RFF survey of the confirmed cases in Luzerne County so that we could obtain information on the costs of illness. We also received help and advice from staff of the Pennsylvania Department of Environmental Resources; in particular, William Cook, acting head, Department of Policy and Planning, in Harrisburg, and Mark Carmon and Eugene Borofski of the Wilkes-Barre regional office. The Department of Environmental Resources provided a list of business establishments affected by the boil-water order and helped us understand the complicated series of events that characterized the Luzerne County outbreak.

Other government officials who provided help and insight into the effects of the Luzerne County outbreak included James Siracuse, Luzerne County Emergency Management Agency in Wilkes-Barre, who was responsible for coordinating the distribution of potable water to residents of the affected communities. Carl Rosencrance, chairman of the Public Works Committee in West Pittston Borough, explained to us how his community government responded to the outbreak. Scott J. Rubin, assistant consumer advocate, Office of Consumer Advocate, Office of the Pennsylvania State Attorney General in Harrisburg, explained the actions taken during the outbreak by his office, the Pennsylvania Public Utility Commission, and both chambers of the Pennsylvania State Legislature.

The Pennsylvania Gas and Water Company (PG&W) in Wilkes-Barre gave us much of the basic information used to reconstruct the events of the outbreak: maps of the water supply and distribution system in the Spring Brook–Hillside Service Area, the area affected by the outbreak; the number of accounts on the boil-water advisory and the length of time individuals were on the advisory; and estimates of the size of the population at risk during the outbreak. We benefited greatly from discussions with J. Glenn Gooch, president; Gerald B. Taylor, vice president for operations and engineering; John F. Kell, vice president and comptroller; and Frank J. Lock, vice president for consumer affairs; Joseph Lubinski, engineering department; Robert R. Brittain, Jr., staff attorney; and Joseph Calabro, director of PG&W's Water Quality Laboratory in Scranton. John Kell was especially helpful to us in responding to the many inquiries from RFF during the period between the completion of the report for EPA in 1985 and the preparation of the manuscript for publication by RFF.

Members of the business community in Luzerne County were helpful in explaining how they coped with the outbreak. We wish to thank Ken Getz, assistant administrator of Nesbitt Memorial Hospital; Jo Anne Lockey of the Pennsylvania Dental Association; the owners of the Apple Tree Nursery School and Little People Day-Care Center; and several restaurant managers for their help.

Additional data on household water use and on the financial operations of restaurants and bars in the affected area were needed for the case study. We wish to thank Jon DeBoer, American Water Works Association, Denver, and Betty Hoyt, research manager, National Restaurant Association, Washington, D.C., for their help in providing these data.

Finally, we wish to thank our colleagues at RFF for their support and help throughout the project, particularly during preparation of the report to EPA. Clifford S. Russell, former director of the Quality of the Environment Division, provided valuable guidance on the measurement of social benefits, both in theory and in practice. Former RFF vice president John F. Ahearne provided an in-depth review of an earlier draft of the manuscript. We are grateful to him for his thoughtful comments and suggestions. Tina Libenson, a senior at Connecticut College in New London and a resident of Luzerne County, interned at RFF and worked with us on the case study. Her knowledge of the area affected by the outbreak and her help with the random telephone survey of households in the affected area, with computations for the analysis of losses due to the outbreak, and with the preparation of draft material for some of the chapters are greatly appreciated. Margaret David served as copy editor for the preparation of the EPA report. We are indebted to her for her efforts to ensure accuracy and consistency throughout the report. Cheryl Johnson did the bulk of the typing of the report, with help from Margaret Parr-Recard. Pat Flynn made all the revisions needed to transform the report prepared for EPA into the current book's manuscript, which Martha Gottron edited for publication by RFF; we are grateful to them and to staff project editor Dorothy Sawicki for their efforts to produce a consistent and readable book.

1 / Introduction

Just before Christmas 1983, Pennsylvania's Department of Environmental Resources declared that contaminated drinking water was the cause of an epidemic of giardiasis—a gastroenteric disease—then raging in the Kingston–Pittston area of Luzerne County in northeastern Pennsylvania. The department advised residents to boil tap water before using it. Before the episode was over, this outbreak of giardiasis would become one of the largest yet experienced, with an estimated 6,000 people falling ill, 75,000 people advised to boil tap water for drinking, and a host of businesses prohibited from using unboiled tap water for drinking or food preparation, in some areas for as long as nine months.

The residents of Luzerne County had fallen victim to an ancient affliction that today is usually associated with developing countries. Certainly, a growing number of Americans are concerned with the safety of their drinking water, but their major concern seems to be cancer and other chronic debilitating illnesses, not acute gastrointestinal illness caused by biological agents. In fact, however, outbreaks of acute waterborne disease in the United States have increased dramatically in recent years, from an average of eleven reported outbreaks per year in the 1950s to thirty-three per year in the 1970s and thirty-one per year in the 1980s (Craun, 1990). Moreover, the number of illnesses per outbreak has increased, from 112 in the 1950s to 244 in the 1980s. The explanation

1

for this increase is not well understood, nor is it clear whether the increase is genuine or simply an artifact of better reporting.[1]

In any event, the cause of the Luzerne County outbreak was a protozoan parasite, *Giardia lamblia*, the most commonly identified pathogen causing waterborne disease in the United States. Between 1971 and 1988, at least 103 outbreaks of giardiasis involving more than 25,000 cases were reported to the federal Centers for Disease Control in Atlanta[2] (Craun, 1990). These numbers almost certainly underestimate the incidence of waterborne giardiasis: for about half the outbreaks of waterborne disease, a cause is never established. Furthermore, the reported outbreaks appear to be concentrated in states that have well-developed reporting programs, suggesting that many outbreaks in other states go unreported.

While the outbreak in Luzerne County was one of the largest ever observed, several large cities could be at risk because the water treatment technology used in those cities is not very effective in preventing *Giardia* contamination. Indeed, according to the Environmental Protection Agency (EPA) in 1987, there were fifteen large unfiltered surface water supply systems in the United States each serving more than 100,000 people and accounting for approximately 16 million of the estimated 21 million people in the United States served by systems using unfiltered surface water.[3]

Because *Giardia* is more resistant than most other pathogens to chlorine disinfection, effective prevention of *Giardia* contamination of surface water supplies will usually require removal of particulate matter in the water by coagulation or sedimentation followed by filtration. In water supply systems in which filtration of raw water either is not practiced or is not sufficiently effective, customers face a potential risk of contracting giardiasis.

The risks posed by *Giardia* have not escaped the notice of the Environmental Protection Agency. Under the Safe Drinking Water Act of 1974,[4] EPA is required to establish drinking water regulations to protect the public from known or anticipated adverse health effects related to drinking water, either by setting maximum levels for contaminants or

---

[1]One theory relates the increase in gastrointestinal illness to the increased concern about carcinogens in drinking water. One of the sources of such carcinogens is said to be the chlorine added during water treatment to kill pathogenic bacteria. In response to such fears, free chlorine levels have been reduced in many drinking water systems.

[2]Before 1980, the federal agency was known as the Center for Disease Control; in 1980 it was renamed the Centers for Disease Control (CDC). The current name is used throughout this volume except in citations to pre-1980 CDC publications.

[3]52 Fed. Reg. 42205 (Nov. 3, 1987).

[4]42 U.S.C.A. §300.

by specifying required treatment techniques if the level of contamination cannot be measured economically or technologically.

By 1983, there was general agreement that *Giardia* posed a public health threat, and although no federal regulation for controlling the incidence of waterborne giardiasis had yet been established, the matter was under consideration at EPA. Before proceeding with such regulations, however, EPA's Office of Drinking Water wanted more information on the costs and benefits of mandating specific water treatment technologies for water supply systems of particular sizes and with particular risks of contamination. In support of EPA's needs, Resources for the Future (RFF) undertook the study presented in this volume in order to develop methods for estimating the benefits of avoiding an outbreak of waterborne giardiasis, and it applied those methods in estimating the benefits of avoiding the 1983–1984 outbreak in Luzerne County. The results of the RFF study were used by EPA, together with analyses that the agency itself had made, in developing the proposed rule for *Giardia lamblia*[5] as part of the National Primary Drinking Water Regulations.

In those regulations, EPA specified (as it was required to do under a 1986 amendment to the Safe Drinking Water Act) criteria under which filtration would be mandated for public surface water supply systems. The final version of the regulations designed to prevent *Giardia* contamination in public water supplies was promulgated by EPA in June 1989.[6]

The costs of complying with these regulations are likely to be substantial, especially for water supply systems required to install coagulation and filtration systems. In those cases, EPA estimated that water bills for customers served by systems with more than 10,000 customers would rise by $4 per household per month, while those served by systems with fewer than 500 customers would see their bills rise by $17 to $32 per month.[7] To put this in perspective, the average American family pays $13.33 a month for its water.[8] Compliance with this regulation would raise that bill by 30 percent in a large system and from 125 to 240 percent in a small system. All told, the capital costs of installing the water treatment systems required by the regulations are estimated to approach $2 billion, with annual operating costs of $339 million.

---

[5]52 Fed. Reg. 42178 (Nov. 3, 1987).

[6]54 Fed. Reg. 27486 (June 29, 1989).

[7]52 Fed. Reg. 42178 (Nov. 3, 1987).

[8]American Water Works Association (1985). This figure represents the average monthly residential water bill, determined in a survey of 368 water suppliers across the country. The estimate has been adjusted from 1984 to 1987 dollars using the GNP (gross national product) deflator.

Faced with an increase of this magnitude, customers of small water supply systems might well wonder what they would be getting for the money and whether it would be worth it. According to a strict reading of the federal law, however, these questions are irrelevant. As in other federal legislation with the primary purpose of protecting the public health (such as the Clean Air Act), the Safe Drinking Water Act does not permit actions taken to protect drinking water to be subjected to a benefit–cost test. Still, as a practical matter, the weighing of benefits versus costs is almost inevitable, if not always explicit, for complete protection from every contingency would be prohibitively expensive.

The benefits associated with the regulation consist largely, but not entirely, of avoiding the damages that would be experienced in future outbreaks in the absence of the regulation. (In addition, the benefits from removal of other contaminants jointly with *Giardia* must be considered.) To estimate the benefits of installing water treatment technology at a given location, then, two questions require answers: What is the likelihood that an outbreak will occur in the absence of any protective action? If an outbreak does occur, what will be the value of the consequences?

THE METHODS: DEVELOPMENT AND APPLICATION

The valuation of the consequences of outbreaks of waterborne disease is the primary concern of this book. As mentioned earlier, it presents a set of theoretical and empirical methods for estimating the benefits of avoiding outbreaks and applies those methods to the estimation of the economic damages resulting from the giardiasis outbreak in Luzerne County. These damage estimates are compared with the actual costs of measures taken to prevent future outbreaks in this area, using a variety of assumptions about the likelihood of a future outbreak in the absence of such measures.

To our knowledge, this study is the first comprehensive attempt to estimate the economic damages of such an outbreak; this fact gives the work a significance that extends far beyond consideration of this specific pathogen or this particular location. Although we believe the study will be of interest to anyone concerned with the economics of events adversely affecting public health, the methods developed should be of particular relevance to episodes sharing one or more of the features of the Luzerne County outbreak. The most important of those characteristics are the following: the outbreak involved contamination of surface water; the health endpoint was acute illness (also called episodic illness), not chronic illness or mortality; and "averting behavior" was feasible:

that is, the affected individuals could, at some cost, protect themselves from exposure.

Few off-the-shelf techniques exist for estimating the benefits of avoiding outbreaks of waterborne disease. Much of the research effort of this project was devoted to the development of theoretical and empirical methods of benefit estimation. We believe that the Luzerne County outbreak was particularly well suited to the development of such methods, containing, as it did, most of the elements likely to be present in other large-scale outbreaks. Most outbreaks of waterborne disease are much smaller and, we assume, less complicated than the one in Luzerne County.

A study of an outbreak of a waterborne disease such as giardiasis reveals three categories of benefits that are particularly underdeveloped methodologically. The first is the valuation of acute morbidity. The great majority of health-related benefit studies have been concerned with valuation of reductions in the probability of death by a particular cause within a given time interval. The approaches to valuing morbidity and mortality are quite different. Morbidity can, in principle, be valued directly: it makes sense to consider what an individual would be willing to pay to avoid an incidence of acute illness and perhaps certain chronic illnesses as well. At the same time, morbidity raises difficult issues regarding the valuation of time and the direct disutility of illness. In contrast, mortality cannot be valued in this direct manner because the individual's willingness to pay to avoid certain death is presumably limited only by the resources available. Instead, mortality valuations are inferred from the estimated willingness to pay to reduce the probability of death.

The second underdeveloped benefits category is the averting behavior of individuals—the actions people take to reduce their exposure to environmental contaminants. While averting behavior can arise in a wide variety of contexts, it is particularly important when drinking water contamination is involved, since those affected may go to considerable trouble to secure uncontaminated water. The valuation of averting behavior requires determining the relationship between averting expenditures, which can be observed, and willingness to pay, which cannot.

The third category of benefits is the least easily monetized. It involves the values associated with reducing pain and suffering and avoiding future anxiety. Surely, people are willing to pay considerable sums to avoid pain and suffering, but although ad hoc valuations are a common feature of lawsuits, no empirical estimates firmly grounded in theory have been developed. Likewise, people who experience an episode of this sort involving contaminated water may not fully trust the integrity of their water supply in the future, no matter what assurances they are given. Unfortunately, we have not been able to make much headway in this third category.

As a result, our benefit estimates consist entirely of the valuation of avoiding morbidity and of averting behavior. Such analyses require information not generally collected in the epidemiological surveys made during disease outbreaks. Therefore, an important part of our research involved the design of questionnaires and other survey instruments to obtain the needed information.

A giardiasis outbreak provides a good laboratory to examine such valuation methods because it exhibits few of the complications that make benefit estimation so difficult in other health contexts. Giardiasis is generally an acute, or episodic, rather than a chronic illness, so its effects are more or less limited to a specific period of time. (It can also be a chronic illness if it goes untreated. See discussion in chapter 2.) Valuation of permanent disability is therefore unnecessary. Because the time between exposure and effects is short, the population at risk is easy to identify, being limited for the most part to those living in the area at the time of the outbreak. Furthermore, the link between cause and effect is not in dispute as it is with some hazardous materials. If credible benefit estimates cannot be obtained for an outbreak of an acute illness such as giardiasis, they are even less likely to be obtainable for the more subtle contaminants that produce cancers or other diseases with long latency periods and long-term health effects.

The results of this study are important to an efficiency evaluation of the EPA policy promulgated in the 1989 regulations and, more specifically, to determining whether water treatment in Luzerne County passes a benefit–cost test. Nonetheless, our results alone are not sufficient either for making national policy or for guiding individual community decisions. For those purposes it is necessary not only to know the probability of outbreaks in the absence of the new policy, but also the benefits associated with the likely removal of other pathogens besides *Giardia*. In addition, as already pointed out in this section, our analysis did not consider important classes of benefits, such as the value of preventing pain and suffering. Finally, other criteria for policy evaluation—concern over equity effects and legal issues, for example—may be just as important as considerations of economic efficiency or perhaps even more so.

ORGANIZATION OF THE BOOK

The remainder of the book is organized in three parts. The first part, which includes chapters 2, 3, and 4, describes the methods used to estimate the economic losses associated with an outbreak of giardiasis. Chapter 2 contains background material on the nature of giardiasis and an anatomy of a waterborne disease outbreak. This results in a taxonomy of the effects of an outbreak, which are classified according to the economic sector affected—individuals, businesses, and government agencies. Chapters 3 and 4 discuss in some detail general methods for estimating losses to individuals and businesses, respectively. Losses to government agencies, which are rather small and relatively straightforward to estimate, also are discussed briefly in chapter 4.

The second part of the book, which includes chapters 5 through 8, contains the case study of the Luzerne County outbreak. Chapter 5 presents the chronology of the outbreak, and chapters 6 and 7 discuss estimates of the losses from the outbreak to individuals, businesses, and government agencies. Chapter 8 summarizes and qualifies the estimates of losses and compares the benefits of avoiding future outbreaks with the actual costs of constructing and operating a treatment plant to filter the water supplied to the affected area. The third part, chapter 9, sketches the contributions of this study and discusses research needs.

REFERENCES

American Water Works Association. 1985. *1984 Water Utility Operating Data*, (Denver, Colo.).

Craun, G. F. 1990. "Causes of Waterborne Outbreaks in the United States." Paper presented at the annual meeting of the American Water Works Association, San Diego, Calif., October 11.

Part 1 / Methods

Part I / Methods

2 / The Nature of Giardiasis and of Community Outbreaks

The most commonly identified pathogen in outbreaks of waterborne disease in the United States today is *Giardia* (Craun, 1979). Waterborne giardiasis has become a major concern to water supply utilities and public health officials across the country, and an understanding of the nature of the disease and its effect on the communities it attacks is helpful in analyzing the costs and benefits of controlling it.

THE NATURE OF GIARDIASIS

Although seldom fatal, giardiasis can be unpleasant. The acute symptoms of giardiasis mimic those of amoebic dysentery, bacillary dysentery, bacterial food poisoning, and "travelers' diarrhea" caused by *Escherichia coli*: explosive diarrhea, marked abdominal cramps, fatigue, weight loss, flatulence, belching, anorexia, nausea, and vomiting (Wolfe, 1979). In a few rare cases, hospitalization for dehydration may be necessary. Ingestion of as few as ten organisms by susceptible individuals may cause acute symptoms, which appear after an incubation period of six to twenty-eight days (Rendtorff, 1979).

Although the acute stage generally lasts only three to four days, untreated giardiasis often develops into a chronic infection, characterized by recurrent periods of acute illness lasting several days. This stage may last for months, leading to malabsorption, increasing debility, and significant weight loss (Craun, 1979). Indeed, it is the intermittent nature of giardiasis, together with the patient's failure to develop a fever, that distinguishes giardiasis from many other gastroenteric illnesses.

The pathogenicity of *Giardia* was first demonstrated in a study around 1950 using prison volunteers (Rendtorff, 1979). A few years later, in 1955, *Giardia* was the suspected cause of a massive epidemic (50,000 cases) of acute gastroenteritis in Portland, Oregon (Veazie, Brownlee, and Sears, 1979). At the time, no cause was identified with certainty. Despite these reports, *Giardia* was regarded in the United States as a generally benign organism until the mid-1960s, when it became evident that giardiasis was one of the most common ailments suffered by Peace Corps volunteers. It was also found to be common among tourists and diplomatic personnel returning from overseas, especially from Leningrad (Wolfe, 1979).

Susceptibility to giardiasis varies considerably in the population, and some people infected with the parasite are asymptomatic. The percentage of adults in the United States infected with *Giardia* has ranged from 2 to 15 percent, according to various random stool surveys discussed by Healy (1979). Evidence suggests that resistance to acute symptoms can be acquired through repeated exposure. A survey of outbreaks of giardiasis in Colorado revealed a lower rate of illness among longterm residents than among recent arrivals to the towns involved. Giardiasis is thought to be an endemic condition in these communities. One explanation for the lower illness rate is that the residents of these communities have become acclimated to giardiasis through repeated exposure to *Giardia* (Wright and coauthors, 1977).

Scarcely twenty years ago, giardiasis was virtually unknown in the United States. While not yet a household word, it is today the most frequent cause of waterborne disease outbreaks for which a cause is identified and a scourge of day-care centers and backpackers. Is this sudden notoriety simply a result of better diagnosis and reporting? Or does it reflect a real increase, for whatever reason, in the incidence of this disease? Whatever the explanation, giardiasis has come to be recognized as a fairly serious public health problem.

Diagnosis and Treatment

Giardiasis is rather difficult to diagnose, in part because the organism is difficult to detect. Unlike many common pathogenic bacteria, *Giardia* cannot be cultured. And while an animal infectivity model is now available, it is not used for routine diagnosis. Instead, testing for *Giardia* requires the search for individual organisms under a microscope.

The most common test to confirm suspected giardiasis is stool sampling. This method, however, is hardly foolproof, with a false-negative rate of 25 to 50 percent when one sample is examined, 10 to 20 percent with two samples, and approximately 1 to 5 percent with three samples (Wolfe, 1979). There are several explanations for the variability of stool test results. First, due to irregular periods of active multiplication, *Giardia* cysts are shed intermittently. Second, many laboratories are not staffed with technicians experienced in the techniques for examining stools for parasites, compounding the inherent difficulties of isolating and visually detecting the cyst in a given sample. Third, specimens are often taken at home by the patient using a stool sample kit, a procedure that may foil proper sampling.[1]

A "string test" may also be used to detect *Giardia* cysts. This technique requires swallowing a capsule attached to a string, withdrawing it after four hours, smearing the bile-stained mucus on a slide, and examining it microscopically. A duodenal aspirate can also be used: a tube is passed through the nose to the stomach and then to the small intestine. These procedures can be quite discomforting for the patient and so they are rarely used.

Because *Giardia* was not considered a problem in the United States until the mid-1960s, it was rarely searched for as a cause of disease. Therefore, little evidence accumulated on its pathogenicity. Because many physicians were not familiar with the symptoms of giardiasis, it is likely that the disease was often misdiagnosed in the past and that inappropriate, perhaps even harmful, treatment was prescribed. Numerous cases of giardiasis lasting intermittently for well over a year have been noted. In one extreme case, a woman with an ileostomy traveled to the Soviet Union and apparently contracted giardiasis in Leningrad. Upon her return she experienced severe gastroenteritis, which was assumed to be a complication of the ileostomy. Two additional operations were performed before giardiasis was suspected, whereupon her symptoms responded rapidly to drug treatment (Schultz, 1979). As giardiasis has become more widely known, the chances of misdiagnosis have declined. Still, even in mountain communities in Colorado, where the disease is well known among doctors, tourists may contract it and return home to areas where it is still relatively unknown. For such patients the probability of misdiagnosis is high.

Once giardiasis is confirmed or strongly suspected, treatment is readily available in the form of several antibiotic drugs. Quinacrine is often recommended as the treatment of choice for giardiasis. Experience has shown quinacrine to have a cure rate of between 85 and 95 percent

---

[1]Personal communication from Pamela Shillam, Colorado Department of Health, Denver, Colo., February 7, 1984. An anonymous reviewer of the manuscript observed that this problem may be disappearing, for a new stool sampling kit has been developed that works well.

(Wolfe, 1979). Possible side effects include intestinal upset, mild headaches, and dizziness. In some ways, quinacrine tends to be a "symptom intensifier," making the cure sometimes worse than the original illness.[2] Incidences of toxic psychosis in adults have also been observed (Wolfe, 1979). A course of the drug runs one week for an adult, with a dosage of 100 milligrams three times daily.

Metronidazole is also frequently prescribed for giardiasis. It has a recorded cure rate of 75 to 85 percent. Side effects may include nausea, headaches, a metallic taste, dizziness, and urine discoloration. Although the Food and Drug Administration has not yet approved metronidazole for treatment of giardiasis, it is much better known and more readily available than quinacrine, since it is often prescribed for other ailments. Metronidazole has been found to be carcinogenic in laboratory mice and mutagenic in bacteria (Dembert, 1981). These circumstances have raised some controversy over its continued use for treating giardiasis, particularly in young children and acutely symptomatic pregnant women.

Furazolidone, a third drug, was not widely used in the United States until the recent increase in giardiasis incidence in day-care centers. It is available in suspension form, making it advantageous for treatment of young children. In a study of thirty-one children aged 10 and under, a cure rate of 77 percent was obtained (Wolfe, 1979); side effects included vomiting, nausea, and diarrhea, fever, and skin rashes. It has also been implicated in some animal tumor studies (Dembert, 1981).

None of the three drugs described above has been explicitly proven safe during pregnancy. When treatment is deemed necessary for pregnant women with confirmed giardiasis, quinacrine is most often chosen. A newer drug, paromomycin, is less efficacious but also less risky for pregnant women, although it can also cause additional nausea and vomiting.

Over-the-counter drugs, especially those for the relief of nausea and diarrhea such as Pepto-Bismol and Kaopectate, may also be used to ease symptoms of giardiasis. Currently, there is no drug commercially available for the prevention of the disease.

Transmission of Giardiasis

Giardia is a protozoan parasite that inhabits the upper portion of the small intestine in mammals, although in cyst form it can survive for several months outside the body. There are two principal stages in its life cycle: the trophozoite and the cyst. The trophozoite stage is the "active" stage, in which the motile protozoa reproduce by binary fission

---

[2]Personal communication from Barry Gaspard, formerly of the Colorado Department of Health, Denver, Colo., February 7, 1984.

and feed in the gut of their mammalian hosts. The cyst is immotile and does not feed. It is the infective stage by which the protozoa are transmitted from one mammal to another.

Giardia cysts are much more resistant to chlorine than either viruses or bacteria are. They are especially resistant in cold water, where they can survive for two months and where the concentration of free chlorine must be much higher than is practical for water treatment (Hoff, 1979).

Exposure to Giardia can come about through contact with the stool of infected animals and people as well as through consumption of contaminated water. While animal hosts include dogs, beavers, muskrats, coyotes, deer, cattle, raccoons, cats, and possibly even birds, the relative importance of these animals in transmission to humans remains unclear (Davies and Hibler, 1979). Studies in Washington state, for example, have shown that pet ownership is a risk factor for giardiasis (Frost and coauthors, no date).

Direct transmission between adults is uncommon but possible; for example, occasional foodborne outbreaks have been observed (Osterholm and coauthors, 1981). On the other hand, the disease has been known to spread rapidly through some institutional populations, such as in mental hospitals and especially in day-care centers. An investigation by the Centers for Disease Control of an outbreak of diarrheal illness in 1975 among young children attending a day-care center in Atlanta found that the introduction of a single Giardia-infected child into the group led to the infection of 54 percent of the other children in the center (Black and coauthors, 1977). The 1-year-olds and 2-year-olds had a higher infestation rate (88 percent) than the 3-year-olds (42 percent), reflecting the inferior personal hygiene habits among the highly mobile but not yet toilet-trained younger children. The significantly higher prevalence of giardiasis in the day-care center children, as opposed to the 2 percent prevalence in age-matched children not in day-care centers, points to the relative ease of child-to-child transmission. Furthermore, person-to-person contact was implicated in the transmission of giardiasis from several infected children to assorted family members. Twenty-five percent of family members, mostly mothers, in contact with a Giardia-positive child were infected with the parasite. In contrast, none of the Giardia-negative children's family members were infected.

Similarly, a 1978 outbreak of giardiasis in a Summit County, Colorado, day-care center involved twenty individuals, both parents and children. In this case, Giardia was brought into the center by the care-giver's sister, who had drunk from a well contaminated by surface runoff. The disease was then spread person to person, first to the children and subsequently to the parents.[3]

---

[3]Personal communication from Shirley Kelly, Summit County Nursing Clinic, Fresno, Colo., February 10, 1984.

For most adults, the biggest danger of exposure to *Giardia* comes from drinking contaminated water. Giardiasis has become a major concern of backpackers and hikers, especially in the mountainous West, but also in the East. Shenandoah National Park, for example, posts signs warning hikers against drinking the *Giardia*-contaminated water. As a result, most backpackers and hikers, not surprisingly, avoid drinking any water that has not been filtered, treated with chemicals, or boiled. Fear of contracting giardiasis has no doubt reduced the pleasures of backpacking and hiking for some people.

The most serious public health problem associated with *Giardia* is its potential for contaminating public water supplies. Humans are thought to be the source of *Giardia* in water supplies, although once a watershed has been contaminated, animals can serve as significant long-term reservoirs of cysts (Davies and Hibler, 1979). Consequently, some watersheds may remain contaminated. *Giardia* cysts in human and animal feces deposited in upland watersheds eventually find their way to watercourses, where they are transported to water supply intakes. Once in the water distribution system, the cysts can be consumed by humans. Unless adequate water treatment is provided, the cycle of contamination involving animals, water supplies, and humans will continue.

Removal of *Giardia* cysts from raw water requires a well-designed and well-operated water treatment plant. In particular, chlorination without filtration may be insufficient. *Giardia* cysts have high resistance to chlorine, particularly at low temperatures. Sufficiently high concentrations of chlorine for sufficiently long contact times will kill *Giardia* cysts. But the contact times required for water temperatures found in most natural environments can strain the holding capacity of many water supply systems unless chlorine levels are so high that palatability and odor are affected.

Outbreaks of giardiasis have occurred in water supply systems with filtration. Outbreaks in Camas, Washington, and Berlin, New Hampshire, occurred in filtered systems, although operating deficiencies were noted in both systems (Kirner, Littler, and Angelo, 1978; Lippy, 1978). The technology required to prevent viable *Giardia* cysts from entering finished water supplies appears to consist of the following steps: chemical pretreatment, which causes solid particles in the raw water to coagulate into larger particles that can be more easily removed by filters; filtration; and disinfection, usually by addition of chlorine (Logsdon, Symons, and Haye, 1979). This treatment train is designated by the Environmental Protection Agency as "conventional filtration treatment." Research shows that diatomaceous earth filtration systems can also be effective (Lange, Bellamy, and Hendricks, 1984).

These assessments of the effectiveness of various water treatment technologies were supported by a recent EPA-sponsored survey of drinking

Table 2-1.   Detections of *Giardia* Cysts in Finished Drinking Water Supplies of the United States, Based on Data Collected 1979–1986 by Hibler (1987)

Classification	Number of sites	Number of positive sites[a]	Percent of sites found positive
Unfiltered, chlorinated	94	16	17.0
Direct filtration[b]	92	17	18.5
Conventional treatment	86	5	5.8
Slow sand and diatomaceous earth filtration	3	0	0.0
Commercial filters and/or pressure filters	12	2	16.7
Cartridge filters	13	7	53.8
Infiltration galleries	16	5	31.3
Filter type unknown	24	6	25.0

Source: Adapted from table V-2 in *Federal Register* notice of proposed rulemaking: (52 Fed. Reg. 42194 (Nov. 3, 1987)).

[a]At each site a number of samples were collected, and a site was "positive" if *Giardia* cysts were found in at least one sample.

[b]May or may not include coagulation or disinfection. The number of systems that applied coagulant and/or polymer, or whether disinfection was interrupted, could not be determined.

water supplies (Hibler, 1987). The results of this survey, shown in table 2-1, indicate that *Giardia* cysts are fairly common even in finished water supplies, suggesting that many treatment systems currently in place do not adequately protect against *Giardia* contamination.[4] Among widely used technologies, conventional treatment is by far the most effective. *Giardia* cysts were detected in 6 percent of the sites employing conventional treatment, but were found in almost 20 percent of sites using disinfection without filtration or "direct filtration" (this category included sites where the treatment technology was not completely characterized by the survey). No cysts were found in systems using diatomaceous earth filtration or slow sand filtration, but the sample was very small.

Based largely on these results, EPA in June 1989 promulgated drinking water regulations designed to prevent *Giardia* contamination.[5] These regulations require all public drinking water systems using surface water sources to have a treatment system capable of removing 99.9 percent of *Giardia* cysts from raw water. Compliance with this rule would be presumed in conventional treatment systems. Direct filtration, slow sand filtration, and diatomaceous earth filtration are also permitted under certain water quality conditions if disinfection is also employed. Fur-

---

[4]Raw water sources were also surveyed, and, as expected, *Giardia* cysts were found in a high percentage of surface waters sampled, but were much less common in groundwater sources.

[5]54 Fed. Reg. 27486 (June 29, 1989).

thermore, all water treatment systems are required to use filtration unless the quality of the raw water is extremely high and the watershed is well protected from human activity.

Many water supply systems, especially in smaller communities, cannot meet these requirements. Ironically, the water systems most at risk tend to be those supplies with otherwise high-quality raw water, free of turbidity and color, for which filtration has never been considered necessary. In rural areas, approximately 3 million households receive drinking water from surface supplies, of which only 3 percent have been pretreated and filtered (Francis and coauthors, no date).

It should not be assumed that waterborne giardiasis is a potential problem only for small communities. As mentioned earlier, during the process of writing the new drinking water regulations, EPA found fifteen water supply systems, each serving more than 100,000 customers, that do not filter the raw water. The population served by these systems exceeds 16 million, or 75 percent of the total of 21.4 million people served by systems using unfiltered surface water (see 52 Fed.Reg. 42205 [1987]).

OUTBREAKS OF WATERBORNE GIARDIASIS IN THE UNITED STATES

The first reported outbreak of giardiasis in the United States (not counting the suspected Portland outbreak) occurred in Aspen, Colorado, in 1965. Since then, there have been more than 100 outbreaks and more than 25,000 illnesses reported (Craun, 1990). Both the number of outbreaks and the number of illnesses are almost certainly large underestimates. After all, a cause is established for only about half of all reported waterborne disease outbreaks (Centers for Disease Control, 1982). Moreover, 44 percent of all reported giardiasis outbreaks have occurred in only one state, Colorado. This uneven geographic pattern suggests underreporting in the other states. Of the reported outbreaks, 72 percent involved surface water supplies. Only 13 percent involved groundwater sources, and the sources of water associated with the remaining outbreaks were not reported. Twenty-one outbreaks involved untreated drinking water; twenty-six, deficiencies in the water distribution or treatment systems. In two of the latter, the treatment was limited to chlorination. Characteristics of some of these outbreaks are shown in table 2-2.

Outbreaks of giardiasis have been reported in every month of the year, but most frequently in June and September. Of fifty-four outbreaks reported between 1965 and 1982, nine occurred in winter, twelve in spring, twenty-one in summer, and twelve in fall (Centers for Disease

Table 2-2. Characteristics of Some Outbreaks of Giardiasis in the United States, 1965 Through 1983

Location of outbreak	Month/year began	Duration	Confirmed cases (no.)	Clinical case definition	Attack rate (%)[a]	Water treatment	Water system failure[b]	Temporary measure taken[b]	Boil-water advisory/order[b]
Aspen, Colo.	12/65	4 mo.	59	Diarrhea for 10 or more days	11.3	Chlorination	Y; distribution system	N.A.	N
Boulder, Colo.	5/72	3 mo.	297	N.A.	33.0	N.A.	N.A.	N.A.	N.A.
Uinta Mountains, Utah[c]	9/74	N.A.	34	Diarrhea for 5 or more days plus other giardiasis symptoms	65.0	None (used natural streams)	N.A.	N.A.	N.A.
Rome, N.Y.	11/74	7 mo.	359	Diarrhea for 5 or more days	10.6	Chlorination	N	N.A.	Y
Estes Park, Colo.[d]	6/75	2 mo.	N.A.	Giardiasis symptoms	37.0	Chlorination	N	N.A.	N
Camas, Wash.	4/76	2 mo.	127	Diarrhea for 10 or more days	4.0	Pretreatment, filtration, chlorination	Y; media loss	Switch to wells	Y
Berlin, N.H.	4/77	2 mo.	275	Diarrhea for 7 or more days	5.0	Flocculation, filtration, chlorination	Y; leaky filters, bad floc	Hyperchlorination	Y
Vail, Colo.	3/78	2 mo.	≥38	N.A.	6.0	Chlorination distribution system	Y	N.A.	N
Leavenworth, Wash.	1/80	5 mo.	17	Diarrhea for 7 or more days	27.0	Filtration and chlorination	Y; significant media loss in filters and insufficient chlorination	Increased chlorine levels	Y

(continued)

19

Table 2-2 (continued)

Location of outbreak	Month/year began	Duration	Confirmed cases (no.)	Clinical case definition	Attack rate (%)[a]	Water treatment	Water system failure[b]	Temporary measure taken[b]	Boil-water advisory/order[b]
Empire, Colo.	8/81	23 days	N.A.	Diarrhea for 5 or more days, or confirmed Giardia in stool	24.0	Chlorination	N	Increased chlorine levels	N
Aspen Highlands, Colo.	10/81	1 mo.	N.A.	Diarrhea for 7 or more days	18.0	Dual-media filter, chlorination	Y; reduced rate of chlorination	N	N
Pagosa Springs, Colo.	12/81	50 days	N.A.	Diarrhea for 3 days with 2 or more giardiasis symptoms	10.0	Sedimentation, sand filtration, chlorination	N.A.	Y; placed Giardia filters on raw and finished water	N
Winter Park, Colo.	1/82	N.A.	N.A.	Diarrhea for 5 or more days	N.A.	N.A.	N.A.	N.A.	N.A.
Reno, Nev.	7/82	6 mo.	300	N.A.	N.A.	No filtration	N.A.	N.A.	Y
Larimar County, Colo. (camp)	9/82	33 days	N.A.	Diarrhea for 5 days, or less than 5 days with gastrointestinal symptoms	28.0	Cannister-type filter system with chlorination	Y; operator did not use filters	N.A.	N.A.
Bailey, Colo.(a)	1/83	4 mo.	4	Diarrhea for 3 or more days plus giardiasis symptoms	11.0	Filtration, chlorination	Y; negative pressure in distribution system	Increased chlorine levels	Y

Mantee, Utah	1/83	1.5 mo.	11	Diarrhea for 2 or more days plus giardiasis symptoms	56.0	Chlorination	N		N
Evergreen, Colo.	6/83	2 mo.	7	Diarrhea for 5 or more days	6.7	Pretreatment, filtration, chlorination	Y; inadequate floc formation (pretreatment)	N.A.	N.A.
Bailey, Colo.(b)	10/83	1 mo.	5	Diarrhea for 4 or more days	27.0	Filtration (cartridge)	Y; cross-connection in distribution system	N.A.	N.A.
Warrior's Mark, Colo.	10/83	2 mo.	10	Diarrhea for 3 or more days	17.0	Filtration, chlorination	N	Increased chlorine levels	Y (or use hot tap water)
Essex Center, Vt.	11/83	5 mo.	20	Giardiasis symptoms	N.A.	Chlorination	N (but inadequate chlorine level)	Changed to alternate, filtered water supply	Y
Danville, Vt.	12/83	5 mo.	20	N.A.	N.A.	Chlorination	N (but zero chlorine residual)	Increased chlorine levels	Y

Sources: Craun (1979); Frost and coauthors (no date); Hopkins and coauthors (1984); Kirner, Littler, and Angelo (1978); and personal communications from Diane Hinton, Utah Department of Health, Summer 1983; Edwin Lippy, Health Effects Research Laboratory, U.S. Environmental Protection Agency, Cincinnati, Ohio. May 1982; Pamela Shillam, Colorado Department of Health, Denver, Colo., February 1984.

[a]The attack rate is defined as the percentage of a random sample with clinical symptoms of giardiasis.

[b]Y = yes; N = no; N.A. = not available.

[c]A group of fifty-two campers.

[d]A group of forty-eight campers.

Control, 1982, 1983). Thus, the reported outbreaks are distributed throughout the year, with those in the summer being a little more frequent and those in the winter being a little less frequent than the average.

While most giardiasis outbreaks have involved small water supply systems (80 percent with fewer than 200 illnesses), most of the reported illnesses through 1982 resulted from four large outbreaks: Rome, New York (1974—4,800 cases estimated); Vail, Colorado (1978—5,000 cases); Lake Havasu, Arizona (1979—2,000 cases), and Bradford, Pennsylvania (1979—3,500 cases). These figures do not include large-scale outbreaks in Reno, Nevada, and Luzerne County, Pennsylvania, in 1983 and 1984, each one of which involved more than 5,000 cases.

Our research produced a number of reports on outbreaks not noted by the EPA, and we have listed these, together with the more prominent outbreaks, in table 2-2. The salient characteristics of each outbreak are also provided. We do not know if these are representative of unreported outbreaks. Nonetheless, these cases serve as a helpful guide to the nature of a giardiasis outbreak.

FOUR STAGES OF AN OUTBREAK

The response to an outbreak of waterborne disease can be divided into the four stages discussed next: discovery, survey and testing, reaction, and aftermath.

The Discovery Stage

Health care providers may be used to seeing occasional cases of giardiasis in their communities. However, when the number of new cases suddenly increases above the normal rate, these people may notify the local or state authorities responsible for investigating disease outbreaks. Reports on elevated disease rates are transmitted in different ways, depending on the locality and the particular state involved. Most often, local physicians report an outbreak directly to state authorities. Surprisingly, at least two outbreaks—those in Vail and in Estes Park, Colorado—were discovered by out-of-state physicians who had treated patients returning from ski vacations.

State, local, or hospital laboratory technicians may also detect *Giardia* organisms in stool samples and alert the authorities. In Colorado, for example, telephone calls from ill individuals seeking advice and requesting stool sampling kits have alerted state health department officials. In Scranton, Pennsylvania, a woman living near an outbreak area but drawing water from a supposedly uncontaminated water supply system had her water tested as a precautionary measure, and *Giardia* cysts were found.

The Survey and Testing Stage

Once several confirmed cases of giardiasis have been discovered in an area, those responsible for monitoring and containing the outbreak move into action. A battery of epidemiological surveys may be administered, and the area's water supply may be tested for the presence of *Giardia* cysts.

Epidemiology   State or local epidemiologists, the federal Centers for Disease Control, or EPA personnel conduct epidemiological surveys to confirm the presence of an outbreak, isolate its cause, and ascertain its size. The testing stage of the outbreak in Berlin, New Hampshire, was particularly thorough. Questionnaires were mailed to all residents with confirmed infections, to a random sample of Berlin residents, and to a random sample of residents from a nearby community (the latter serving as a control group). A random stool survey was also instituted, and hospital emergency room records were searched for patients admitted with symptoms of gastroenteritis. When an outbreak is suspected in a tourist town, even greater efforts to determine the cause may be made. During the outbreaks in Aspen, Vail, and Winter Park, Colorado, state health officials contacted tour operators, lodge personnel and guests, ski clubs, and other groups to track down tourists exposed to *Giardia* cysts during their vacations.

Informal methods may also be used to isolate the cause of an outbreak and roughly gauge its size. Pharmacists may be asked about drug purchases, doctors about complaints of gastroenteritis, and schools about absenteeism. These methods are less effective, primarily because the effects of the outbreak are often hard to distinguish from disease caused by other events in the community.

In describing the size and other characteristics of an outbreak, epidemiologists distinguish among three kinds of cases:

1. *Confirmed cases*—persons for whom *Giardia* is found in stool samples.
2. *Clinical cases*—persons who report symptoms of giardiasis without positive stool samples. This category can include symptomatic individuals who have negative stool samples and individuals who have not had stool samples taken.
3. *Asymptomatic cases*—persons who have positive stool samples but no symptoms.

These designations are independent of whether the individual has seen a doctor. For example, a confirmed case may be established by a stool sample conducted in a doctor's office or by a self-administered stool sample kit (the individual must send the sample to a laboratory). Likewise, a clinical case may refer to anyone whose symptoms meet a clinical case definition, whether diagnosed in a doctor's office or identified in a response to a random epidemiological survey.

The epidemiological investigation generally includes both a survey of the confirmed cases and a questionnaire sent to a random sample of the exposed population. The survey of confirmed cases is used to generate hypotheses about the cause of the outbreak. For example, the people surveyed are asked about the source of drinking water in the home, their exposure to children in day-care centers, whether they have pets, and whether they have recently drunk unpurified water while hiking. Information on other risk factors, such as age, sex, years of residence, and quantity of water consumed, is also collected.

If considered desirable, and especially if a local public water supply is implicated, these hypotheses are tested in a random sample of residents. Random sample surveys ask for similar, but often less-detailed, information. Unlike confirmed-case surveys, these surveys ask respondents if their symptoms match a clinical case definition. Positive responses to this question are often called clinical cases even though the respondent has not consulted a physician.[6] If contamination of a water supply is suspected, a control area is also sampled, and clinical illness rates are calculated for both the suspected outbreak area and the control area. Evidence of contamination is provided if the difference between the two illness rates is statistically significant.

In Camas, Washington, for example, water was suspected largely because no other common risk factor was found among the confirmed cases. The municipal water in Camas comes from two different types of sources—two mountain streams and seven deep wells. Surveys of users implicated the city's surface water supply as the most likely source of the organism: six of the thirty-six individuals using surface water showed symptoms of giardiasis lasting seven days or more, while none of the forty users of well water surveyed showed any symptoms of the illness.

If a random survey is conducted, the illness rate can also be used to estimate the size of the outbreak, although that is not its primary pur-

---

[6]By chance, some respondents may have confirmed cases. Thus, in this study we define *attack rate* to be the sum of clinical and confirmed cases in the sample divided by the number sampled. We also distinguish between this *gross attack rate* and the *net attack rate*. The net attack rate is the gross rate minus the background rate—the background rate being the attack rate in areas where no outbreak is suspected (a control area, or the outbreak area prior to the outbreak period).

pose. The illness rate is multiplied by the population served by the contaminated water supply. (Sometimes the illness rate in the control area is considered the "endemic," or "background," rate prevailing prior to the outbreak and is subtracted from the rate in the affected area to compute an illness rate attributable to the outbreak.)[7]

The results of epidemiological studies (Lopez and coauthors, 1980) undertaken during the outbreak in Berlin, New Hampshire, in 1977 can be used to illustrate the relationship between confirmed and clinical cases. Of the 275 people with confirmed cases of giardiasis reported by the local hospital, 90 percent were Berlin residents. The other 10 percent either worked in Berlin or visited the city frequently. During the height of the outbreak, the Centers for Disease Control surveyed a random sample of approximately 700 Berlin residents and approximately 700 residents in two control towns. For this survey, a clinical case of giardiasis was defined as any diarrheal illness lasting seven days or longer. About 5 percent of the Berlin residents suffered an illness consistent with that definition of giardiasis. In contrast, the clinical rate in the two other towns was less than 2.5 percent. Using an illness rate of 5 percent, it was estimated that approximately 750 of Berlin's 15,000 residents were affected. However, a community stool survey revealed that 47 percent of the population was infected with *Giardia* during the epidemic, although the majority of these infected individuals were asymptomatic. The reason for the high rate of asymptomatic infection is unknown. A follow-up stool survey done nine months later showed a lower infection rate (3.4 percent), indicating that the epidemic had ended.

The experience in Berlin illustrates that the reported number of cases must be used with care. Often what is reported is the number of confirmed cases, an estimate consisting of those individuals who test positive for *Giardia*. Since not every ill person sees a doctor or complains to public health authorities and since *Giardia* sampling has a fairly high false-negative rate, the number of confirmed cases is usually a substantial underestimate of the size of the outbreak.

The results of a population survey—the number of individuals reporting symptoms of illness—are more likely to generate a useful estimate of the size of the outbreak. Even these estimates are subject to more than just random error. For one thing, the clinical definition of giardiasis in most studies is an attack of diarrhea that lasts a specified number of days. Thus, diarrhea due to other causes may be attributed

---

[7]These random sample studies often determine more than the extent and cause of an outbreak. Associations between length of residence and disease incidence and between number of glasses of water drunk daily and disease incidence are two prominent findings in unpublished epidemiological studies of outbreaks conducted by the Colorado Department of Health. These studies are summarized in Hopkins and coauthors (1984).

to giardiasis and giardiasis attacks of shorter duration may be ignored. Moreover, the particular characteristics of two different outbreaks may dictate the use of two different case definitions, making comparison of size of the outbreaks difficult. For example, the duration of symptoms used to define a clinical case of giardiasis was five days for Rome, New York; seven days for Berlin, New Hampshire; and ten days for Camas, Washington. Reported clinical cases are neither comparable nor additive.

In addition, the survey data give little information on the severity of the disease. For the great majority of survey cases reported, a physician was not consulted. While people may have many reasons for not consulting a physician, surely one of the most common is simply that the illness is not very serious. All reports of symptoms in the literature are from physicians about their patients; a fortiori, these are biased toward the more serious cases. On the other hand, after an outbreak is publicized, people with relatively minor symptoms, and even those without symptoms but who believe they have been exposed, may seek medical attention.

Water System Testing As the epidemiologists become more suspicious of the local water supply, the raw and finished water supplies will be tested, the water treatment and distribution systems inspected, and sometimes the watershed surveyed for possible sources of contamination.

Water distribution systems in outbreak areas are most often supplied by surface water or infiltration galleries.[8] Treatment is generally limited to chlorination, although filtration without chemical pretreatment may also be provided. Neither treatment method is completely effective in removing or killing *Giardia* cysts. The rudimentary nature of this type of water treatment generally reflects an otherwise high-quality water supply. It is no accident that most communities experiencing giardiasis outbreaks are located in the mountains and enjoy high-quality raw water that meets federal turbidity and coliform standards for municipal drinking water.

Water supply systems with more advanced treatment may also be at risk if monitoring and maintenance are inadequate. For example, the Camas system used chemical pretreatment, filtration, and chlorination. Nonetheless, inspection of the system revealed significant deterioration of the plant's filters as well as inadequate pretreatment. The combination

---

[8]An infiltration gallery is a shallow trench dug parallel to the stream to collect water percolating through the streambed.

of these defects allowed *Giardia* cysts to pass through the treatment system. The EPA found cysts in both the raw and filtered water supplies.

Watershed investigations have often involved searching for and testing beavers for *Giardia*. In Camas three beavers found within foraging distance of the water intakes tested positive for *Giardia* cysts. Camas thus became the first outbreak in which the beaver was implicated as an effective animal reservoir for *Giardia* cysts and as the most likely source of water contamination. In Berlin investigators found a beaver lodge in one of the storage reservoirs. An autopsy found *Giardia* trophozoites in one of the four captured beavers. Of course, beavers are not the only possible source. A survey taken of homes and institutions located upstream from one of the treatment plants in Berlin revealed more than forty sanitation violations.

Even sophisticated water treatment systems may be at risk. Evergreen, Colorado, for example, experienced a small outbreak of giardiasis in 1983. It was concluded that the water supply system was the source of the illness, even though the water had been chemically pretreated, put through a rapid-sand filter, and chlorinated. However, the plant experienced difficulties in achieving adequate floc formation and also had had problems with low alkalinity. Sudden changes in raw water flow and turbidity may have resulted in inadequate chemical pretreatment.[9]

The Reaction Stage

Once the water supply has been implicated, local authorities and the water utility enter the reaction stage. Local authorities may issue boil-water advisories, and perhaps even orders, to minimize public exposure. In addition, temporary measures may be taken to repair the water supply system, decontaminate the water supply, and reduce risks of recontamination.

The nature of boil-water advisories and orders varies from outbreak to outbreak, probably because agreement is lacking on effective precautions. Some states and localities simply advise residents to forgo use of tap water, others counsel boosting temperatures of hot water heaters, and still others recommend boiling the contaminated water for one minute. While drinking water is always included in advisories, Charleston, New Hampshire, among other outbreak sites, added bathing and cooking to its list of affected water uses (*Union Leader*, 1979).

---

[9]Personal communication from Joseph Meek, Colorado Department of Health, Denver, Colo., February 8, 1984. Recall that Berlin, New Hampshire, and Camas, Washington, also had operating filter systems at the time of their outbreaks.

Because individuals place only themselves and their families at risk by drinking contaminated water, they cannot be ordered to avoid the untreated tap water. In one case, authorities in Berlin felt that their advisory was not being taken seriously; they reissued the advisory, this time with more media coverage. Nonetheless, it is generally thought that compliance with boil-water advisories during giardiasis outbreaks is high because the number of new cases reported usually falls off rapidly after the boil-water advisory is issued.

A boil-water order may be issued for businesses and government offices. Many businesses can be affected, but little is known about the actual steps taken to comply with such an order. Boil-water orders trigger government surveillance and enforcement activities, and government agencies may also provide uncontaminated water to the affected communities. For the giardiasis outbreak in Tooele, Utah, in 1983 (1,272 cases), the state health department estimated administrative costs to be $11,000 for the health department and $16,000 for the local government (Newman, 1983).

How the water utility reacts depends partly on the design of the water treatment and distribution systems and partly on whether the systems were operating properly at the time of the outbreak. In Camas, clean well water used by part of the system was pumped to all residents (Kirner, Littler, and Angelo, 1978). In Bailey, Colorado, special 1-micron filters were installed to trap the tiny cysts in the raw water.[10] In Warrior's Mark, a subdivision near Breckenridge, Colorado, which uses rapid-sand filtration and chlorination to treat its raw surface water, the residual chlorine level was increased to about 5 milligrams per liter, a level comparable to that in swimming pools, while the contaminated filter media was being replaced.[11] In Luzerne County the chlorine level was also raised temporarily, whereupon stories appeared in the news about taste and odor problems, ruined laundry, and infants breaking out in rashes (*Wilkes-Barre Times Leader*, 1984). Thus, the actions taken by the water utility to reduce exposure to *Giardia* cysts may carry other risks, even as the threat of giardiasis is being reduced.

The Aftermath Stage

With the boil-water advisory or order in place and temporary measures taken to reduce the risk of contamination, the outbreak may be considered stabilized. At this time, discussions begin concerning long-term

[10]Personal communication from Steve Snider, Colorado Department of Health, Denver, Colo., February 1984.

[11]Personal communication from Joseph Meek, Colorado Department of Health, Denver, Colo., February 1984.

solutions to the problem and how the costs of the outbreak and of protection against future outbreaks are to be shared. These deliberations and subsequent decisions are not documented in the typical published epidemiology study, but, as with the Luzerne County outbreak (see the subsection on "The Aftermath" in chapter 5), they may include lawsuits initiated by affected residents or by businesses attempting to recover certain costs of the outbreak from the water utility. This type of activity is a relatively recent aspect of giardiasis outbreaks.

Clearly, the physiological and behavioral consequences of an outbreak of giardiasis have costs, as described in this chapter. The next two chapters develop estimation methods and procedures based on theoretical considerations to measure the economic losses of an outbreak. This process involves abstracting from the details of what happens during an outbreak to a consideration of how people who are trying to maximize personal welfare (in the broadest sense) react when their environment changes. Our goal is the building of economic estimation models that attempt to capture the key features of individual reactions.

REFERENCES

Black, Robert E., Aubert C. Dykes, Susanne P. Sinclair, and Joy G. Wells. 1977. "Giardiasis in Day-Care Centers: Evidence of Person-to-Person Transmission," *Pediatrics* vol. 60, no. 4, pp. 486–491.
Craun, Gunther F. 1979. "Waterborne Outbreaks of Giardiasis," in Walter Jakubowski and John C. Hoff, eds., *Waterborne Transmission of Giardiasis.* Prepared for U.S. Environmental Protection Agency by Health Effects Research Laboratory and Municipal Environmental Research Laboratory, Cincinnati, Ohio, June (Available from National Technical Information Service, Springfield, Va. PB 299-265).
Centers for Disease Control. 1982. *Water-Related Disease Outbreaks Annual Summary 1981*, Health and Human Services Publication (CDC) 82-8385 (Atlanta, Ga., U.S. Department of Health and Human Services).
————. 1983. *Water-Related Disease Outbreaks Annual Summary 1982*, Health and Human Services Publication (CDC) 83-8385 (Atlanta, Ga., U.S. Department of Health and Human Services).
Davies, Robert B., and Charles P. Hibler. 1979. "Animal Reservoirs and Cross-species Transmission of *Giardia*," in Walter Jakubowski and John C. Hoff, eds., *Waterborne Transmission of Giardiasis.* Prepared for U.S. Environmental Protection Agency by Health Effects Research Laboratory and Municipal Environmental Research Laboratory, Cincinnati, Ohio, June (Available from National Technical Information Service, Springfield, Va. PB 299-265).
Dembert, Mark L. 1981. "Giardiasis," *American Family Physician* vol. 23, no. 2, pp. 137–140.

Francis, Joe, Bruce Brower, Wendy Graham, Oscar Larson, Julian McCaull, and Helene Vigorita. No date. *National Statistical Assessment of Rural Water Conditions* vol. 3. Prepared for Office of Drinking Water (Washington, D.C., U.S. Environmental Protection Agency).

Frost, Floyd, Lucy Harter, Byron Plan, Karen Fukataki, and Bob Holman. No date. "Giardiasis in Washington State." Environmental Health Surveillance, Office of Environmental Health Programs, Health Division, Washington State Department of Social and Health Services. Unpublished.

Healy, G. R. 1979. "Studies on Parasite Prevalence in the U.S.," in Walter Jakubowski and John C. Hoff, eds., *Waterborne Transmission of Giardiasis.* Prepared for U.S. Environmental Protection Agency by Health Effects Research Laboratory and Municipal Environmental Research Laboratory, Cincinnati, Ohio, June (Available from National Technical Information Service, Springfield, Va. PB 299-265).

Hibler, C. P. 1987. *Analysis of Municipal Water Samples for Cysts of Giardia.* Report prepared for Office of Drinking Water (Washington, D.C., U.S. Environmental Protection Agency).

Hoff, J. C. 1979. "Disinfection Resistance of *Giardia* Cysts: Origins of Current Concepts and Research in Progress," in Walter Jakubowski and John C. Hoff, eds., *Waterborne Transmission of Giardiasis.* Prepared for U.S. Environmental Protection Agency by Health Effects Research Laboratory and Municipal Environmental Research Laboratory, Cincinnati, Ohio, June (Available from National Technical Information Service, Springfield, Va. PB 299-265).

Hopkins, Richard S., Pamela Shillam, Barry Gaspard, Linda Eisnach, and Richard J. Karlin. 1984. "Waterborne Disease in Colorado: Three Years' Surveillance and 21 Waterborne Outbreaks." Denver, Colorado, Department of Health. Unpublished.

Kirner, J. C., J. D. Littler, and L. A. Angelo. 1978. "A Waterborne Outbreak of Giardiasis in Camas, Wash.," *American Water Works Association Journal* vol. 70, no. 1, pp. 35–40.

Lange, K. P., W. D. Bellamy, and D. W. Hendricks. 1984. "Filtration of *Giardia* Cysts and Other Substances. V.1. Diatomaceous Earth Filtration." Prepared for U.S. Environmental Protection Agency, Washington, D.C., June (Available from National Technical Information Service, Springfield, Va. PB 84-212703).

Lippy, Edwin C. 1978. "Tracing a Giardiasis Outbreak at Berlin, New Hampshire," *American Water Works Association Journal* vol. 70, no. 9, pp. 512–520.

Logsdon, G. A., J. M. Symons, and R. I. Haye, Jr. 1979. "Waterborne Filtration Techniques for Removal of *Giardia* Cysts and Cyst Models," in Walter Jakubowski and John C. Hoff, eds., *Waterborne Transmission of Giardiasis.* Prepared for U.S. Environmental Protection Agency by Health Effects Research Laboratory and Municipal Environmental Research Laboratory, Cincinnati, Ohio, June (Available from National Technical Information Service, Springfield, Va. PB 299-265).

Lopez, Carlos E., Aubert C. Dykes, Dennis D. Juranek, Susanne P. Sinclair, Judith M. Conn, Robert W. Christie, Edwin C. Lippy, Myron G. Schultz, and Maynard H. Mires. 1980. "Waterborne Giardiasis: A Communitywide Outbreak of Disease and a High Rate of Asymptomatic Infection," *American Journal of Epidemiology* vol. 112, no. 4, pp. 495–507.

Newman, Sumner D. 1983. "Direct Cost Estimates of Giardiasis Outbreak in Tooele." Memorandum to Michael J. Stapley, Acting Executive Director of the Bureau of Management Audit, State of Utah Department of Health, December 21.

Osterholm, Michael T., Jan C. Forfang, Terry L. Ristinen, Andrew G. Dean, John W. Washburn, Janice R. Godes, Richard A. Rude, and John Mc-Cullough. 1981. "An Outbreak of Foodborne Giardiasis," *The New England Journal of Medicine* vol. 304, no. 1, pp. 24–28.

Rendtorff, R. C. 1979. "The Experimental Transmission of *Giardia lamblia* among Volunteer Subjects." Reprinted in Walter Jakubowski and John C. Hoff, eds., *Waterborne Transmission of Giardiasis.* Prepared for U.S. Environmental Protection Agency by Health Effects Research Laboratory and Municipal Environmental Research Laboratory, Cincinnati, Ohio, June (Available from National Technical Information Service, Springfield, Va. PB 299-265).

Schultz, M. 1979. "Discussion on 'Managing the Patient with Giardiasis: Clinical, Diagnostic and Therapeutic Aspects,' " in Walter Jakubowski and John C. Hoff, eds., *Waterborne Transmission of Giardiasis.* Prepared for U.S. Environmental Research Laboratory, Cincinnati, Ohio, June (Available from National Technical Information Service, Springfield, Va. PB 299-265).

Union Leader. 1979. "Charlestown Reservoir Infested by Parasites," Manchester, N.H., Friday, April 13.

Veazie, Lyle, Inez Brownlee, and H. J. Sears. 1979. "An Outbreak of Gastroenteritis Associated with *Giardia lamblia,*" in Walter Jakubowski and John C. Hoff, eds., *Waterborne Transmission of Giardiasis.* Prepared for U.S. Environmental Protection Agency by Health Effects Research Laboratory and Municipal Environmental Research Laboratory, Cincinnati, Ohio, June (Available from National Technical Information Service, Springfield, Va. PB 299-265).

Wilkes-Barre Times Leader. 1984. "PUB Told of Water Problems." Wilkes-Barre, Pa., February 8.

Wolfe, Martin S. 1979. "Managing the Patient with Giardiasis: Clinical, Diagnostic and Therapeutic Aspects," in Walter Jakubowski and John C. Hoff, eds., *Waterborne Transmission of Giardiasis.* Prepared for U.S. Environmental Protection Agency by Health Effects Research Laboratory and Municipal Environmental Research Laboratory, Cincinnati, Ohio, June (Available from National Technical Information Service, Springfield, Va. PB 299-265).

Wright, Richard A., Harrison, C. Spencer, Richard E. Brodsky, and Thomas M. Vernon. 1977. "Giardiasis in Colorado: An Epidemiologic Study," *American Journal of Epidemiology* vol. 105, no. 4, pp. 330–336.

3 / Measuring Losses to Individuals

$\mathbf{A}$n outbreak of waterborne disease has many consequences for the community it affects. Some people become ill. Ill or not, most people take measures to avoid the contaminated water. Businesses, especially those involved in food handling, may be required to find alternative sources of water. Government also has responsibilities: obtaining safe water for schools and hospitals, enforcing water use restraints on businesses, perhaps helping arrange for water distribution to citizens. Governmental agencies often must conduct epidemiological and other investigations in response to elevated incidence of illness. The water supply utility, whether public or private, is also affected. It must eliminate contaminants in the water supply and distribution system, restore public confidence, and perhaps help arrange for alternative water supplies for the duration of the episode.

This list is hardly exhaustive, but it should be enough to suggest the varied and possibly enormous economic consequences of an episode of contaminated drinking water. The economic damages of a water contamination episode are defined here as the sum of the economic consequences sustained by each individual, business, and government directly or indirectly affected by the outbreak. These consequences can be valued using the principles of applied welfare economics. In this chapter and the next, we present a taxonomy of effects of water supply contamination and describe in some detail the methods used to value the specific effects that resulted from the outbreak of giardiasis in

Luzerne County, Pennsylvania, that occurred between December 1983 and September 1984.

Several previous attempts to measure the cost of disease outbreaks have been made. One of the earliest was an estimate of the cost of an epidemic of encephalitis in Dallas, Texas, in 1966 (Schwab, 1968). Levy and McIntire (1974) investigated the economic impact of an outbreak of foodborne salmonellosis in St. Louis, Missouri. Baker and coauthors (1979) studied the economic effects of an outbreak of waterborne gastrointestinal illness in Sewickley, Pennsylvania. That study was noteworthy for its inclusion of some averting expenditures; the researchers included the increase in purchased bottled water as a cost of the outbreak. Each of these studies reported difficulties in estimating the losses to business. In addition, methodological difficulties, notably the confusion of costs, benefits, and damages, may have led to double counting of some effects.

These methodological difficulties, as well as others, have attracted the attention of economists studying economic consequences of natural events such as floods and blizzards. A notable example is the study by Russell, Arey, and Kates (1970) of the damages caused by a drought in Massachusetts in the 1960s. That study's discussion of the problems of estimating "episode" damage was our point of departure for the development of a theoretical framework.

Before turning to specific effects of a disease outbreak, several general points about applied welfare analysis should be made. To begin with, it is important to distinguish between damages (or losses) and benefits. These terms are often used as if one were the mirror image of the other, but, in fact, they are logically and semantically distinct. That is, we speak of the damages or losses from an *event*, such as a flood, and we speak of the benefits of a *policy*. Suppose a flood occurs, and upon calculating its effects, the damages (losses) are found to be X. Suppose further that a dam is proposed to protect against future floods. The dam will presumably offer future benefits in the form of flood damages avoided, but at the time the dam is built, it is impossible to know exactly what those benefits will be. They can be *estimated*, however, provided the future with and without the dam can be predicted. For instance, suppose that a future flood is estimated to cause the same damages as the most recent flood, or X. A study of hydrological records together with the storage capacity of the reservoir concludes that the probability of a flood of a given magnitude in any year is p without the dam and p^* with it. Then, the expected annual benefits of the dam are calculated to be $(p - p^*)X$. But the damages of a future flood are X, as are, in this case, the actual damages of the most recent flood. In this study, we are concerned primarily with the damages of a known outbreak of drinking water contamination, although chapter 8 presents some estimates of benefits.

A second point concerns the distinction between individual and social welfare. By definition, the social benefits associated with a change in the environment are the sum of the amounts each individual is willing to pay for the change (or its avoidance). For most applications it is sufficient to examine the effects on the individuals directly affected by the change. Thus, the evaluation of a new recreation area, say, can often be limited to consideration of changes of consumer surplus among recreation participants.

Evaluation of morbidity (or at least its employment effects) is not quite so simple, both because illness affects the individual's contribution to social welfare through absence from work and because the individual may be compensated for the lost income.

For example, consider a day's work lost due to illness. Apart from other suffering, individuals may or may not incur economic losses equal to their daily wages, depending on the availability of sick leave. But what of the employer's loss of output due to the employee's absence? Although in the long run the marginal value of the employee to the business is at least equal to the wage rate, it may be incorrect to apply this value to determine the impact of short absences. In many cases, short absences of a few employees may have no effect at all on output, because healthy workers work harder to cover for those who are sick or recovered workers work harder to clear up their backlogs after they return to work. If so, the employees' illness damages not the employer, who loses nothing (if sick leave is paid) or even enjoys a benefit (if there is no paid sick leave),[1] but the employees' hardworking coworkers or themselves, once they return to work.

Even if output is lost, firms outside the affected area may be able to make it up. In this case, the employees' illness would not result in national social losses, although it would clearly result in local losses (matched by local benefits elsewhere).

As a practical matter, tracing the incidence of these losses would mire us in a swamp of detail involving production and marketing relationships, as well as institutional considerations such as the availability of paid sick leave or medical insurance. To avoid these complications, the individual is assumed to be self-employed, so that the individual's interest and the employer's interest are identical. The before-tax wage is assumed to be an adequate representation of the social value of lost work. The before-tax stipulation is made because the losses to the individual are based entirely on the after-tax wage. After describing the losses to the indi-

---

[1]Sick leave makes it less expensive for employees to miss work during illness, and, in fact, some studies have shown that sick leave policy can effect absenteeism (Winkler, 1980). Except for this "moral hazard" problem, however, sick leave does not affect the costs of illness but the incidence of those costs.

vidual, we make an adjustment to account for the broader social losses as represented by the lost tax revenue.

Similar assumptions are made in other situations. For example, medical personnel were not asked if they would have been treating other patients in the absence of an episode. It was assumed that they would have and therefore that opportunities were lost.

Finally, many different types of losses are associated with an outbreak of waterborne disease, and we attempt to measure only some of them, including the following: the costs of medical care; the costs of actions taken by individuals and businesses to avoid the disease, such as the costs of bottled water; the costs of water supply system surveys; the costs and damages associated with temporary measures taken to ensure that drinking water supplies are safe; and the costs to government agencies involved in identifying the outbreak and facilitating community response. We do not attempt to measure the important losses associated with pain and suffering, with anxiety over the possibility of contracting the disease, and with diminished intrinsic value resulting from the loss of a "pure" supply of drinking water.

What follows is a more detailed discussion of measuring the effects of a disease outbreak on individuals, developing two economic models from which are derived expressions for the willingness of individuals to pay to avoid using contaminated water. These models guide the empirical work presented in chapter 6. Readers not familiar with formal economic models may wish to skip to the section entitled "Empirical Considerations."

VALUATION OF ILLNESS

This section describes a model of individual utility maximization, from which an expression is derived for the losses due to an episode of environmental pollution such as contaminated drinking water. As noted above, an adjustment will be made to account for social losses rather than losses to the victim of illness.

The model described below is an extension of earlier work by Harrington and Portney (1987). Goods and time are inputs to a household production function producing activities from which people derive utility. The total consumption of goods is constrained by a budget, which in turn depends on the amount of time devoted to work. The time devoted to work and leisure is constrained by the total time available. Contaminated drinking water affects utility in several ways. First, individuals may, at some cost, take averting action, such as purchasing bottled water, to avoid contact with the contaminated water. Second, sick individuals incur medical expenses. Third, if illness does strike, the time available for work or leisure is reduced. Illness is also assumed to affect worker productivity even on days when work is not missed. This element is important in estimating the losses associated with a lingering or intermittent illness.

Suppose, then, that an individual combines leisure time L and expenditure X on goods to produce satisfaction. Suppose also that the individual cannot control the level of contamination C but can fend off some of its effects through defensive expenditures D, as in the following utility function:

$$U^*(X,L,D;C) = F(D,C)\,U(X,L), \tag{3-1}$$

where F is one's "productivity" in producing utility, $0 \le F \le 1$, and is a function of D. We assume U_X, $U_L > 0$; U_{XX}, $U_{LL} < 0$; $F_D > 0$, $F_C < 0$, $F_{DD} < 0$, $F_{CC} < 0$. Because the individual is assumed to be self-employed, this same productivity factor is assumed to affect work performance and therefore wage income. In addition, the contamination could incapacitate the individual for some period of time, making him or her unavailable for either work or leisure. Denote this sick time by $S(D,C)$; as indicated, it is, like F, dependent on the level of defensive expenditures and the contamination level, where $S_D < 0$, $S_C > 0$, $S_{DD} > 0$, $S_{CC} > 0$.

The individual maximizes equation (3-1) subject both to a time constraint

$$L + W + S = T, \tag{3-2}$$

where W is work time, and T is total time available, and to a resource constraint

$$I + wF(D,C)W \ge mS + D + X \tag{3-3}$$

In equation (3-3) medical expenses mS are assumed to be proportional to duration of illness S, I represents nonwage income, and w the rate at which income is produced from working in the absence of contamination-induced illness (which is referred to as the wage rate). In effect, the individual is assumed to engage in piece work and pay is thus adjusted by the productivity factor F. The individual may work as many or as few hours as he or she likes. Thus, the individual maximizes

$$\pounds = F(D,C)U(X,L)$$
$$+ \lambda[I + F(D,C)w(T - L - S) - D - X - mS], \tag{3-4}$$

where the term in brackets is the "full income" constraint of Becker (1965). The first order conditions are

$$\pounds_X = F(D,C)U_X - \lambda = 0 \tag{3-5a}$$

$$\pounds_L = F(D,C)U_L - \lambda wF(D,C) = 0 \tag{3-5b}$$

$$\pounds_D = F_DU + \lambda F_DwW - \lambda FwS_D - \lambda - \lambda mS_D = 0 \tag{3-5c}$$

By considering the amount of additional income required to keep the individual on the same indifference curve, the individual's marginal willingness to pay (*WTP*) to avoid a small increase in contamination can easily be expressed in terms of the derivatives of the indirect utility function *V*.

$$WTP = -\frac{V_C}{V_I} \tag{3-6}$$

By the envelope theorem, the derivatives of the indirect utility function are

$$V_I = \lambda \tag{3-7a}$$

$$V_C = F_CU + \lambda F_CwW - \lambda FwS_C - \lambda mS_C \tag{3-7b}$$

Therefore, willingness to pay can be written

$$WTP = -\left(\frac{F_DU}{\lambda} + F_DwW\right)\frac{F_C}{F_D} + (FwS_D + mS_D)\frac{S_C}{S_D} \tag{3-8}$$

Now consider the total change in sick time and productivity with a change in contamination:

$$\frac{dS}{dC} = S_DD_C + S_C, \quad \frac{dF}{dC} = F_DD_C + F_C$$

or

$$\frac{S_C}{S_D} = \frac{1}{S_D}\frac{dS}{dC} - D_C, \quad \frac{F_C}{F_D} = \frac{1}{F_D}\frac{dF}{dC} - D_C$$

Substituting these expressions into equation (3-8) produces

$$WTP = -\frac{U}{\lambda}\frac{dF}{dC} - wW\frac{dF}{dC} + Fw\frac{dS}{dC} + m\frac{dS}{dC}$$

$$+ \left(\frac{F_DU}{\lambda} + F_DwW - FwS_D - mS_D\right)D_C \tag{3-9}$$

From equation (3-5c), the bracketed expression in equation (3-9) is equal to 1. Also, by differentiating the time constraint (3-2), we have

$$L_C + W_C + dS/dC = 0$$

Therefore,

$$WTP = -\frac{U}{\lambda}\frac{dF}{dC} - wW\frac{dF}{dC} - FwW_C$$

$$- FwL_C + m\frac{dS}{dC} + D_C \qquad (3\text{-}10)$$

As noted above, to obtain the social welfare losses associated with individual illnesses, lost tax revenue associated with lost work must be added to equation (3-10). Hence, if G represents individual income taxes and SW social welfare, then

$$\frac{\partial SW}{\partial C} = WTP - \frac{dG}{dC}$$

(The sign is negative because dG/dC represents the reduction in tax revenues). If w^* is the wage rate before taxes (that is, individual productivity in the absence of illness), then taxes collected can be written as

$$G = F(w^* - w)W$$

so that

$$\frac{dG}{dC} = \frac{dF}{dC}(w^* - w)W + F(w^* - w)W_C$$

The marginal loss of social welfare associated with individual response to increased contamination is therefore

$$\frac{\partial \text{ Social Welfare}}{\partial C} = -\frac{U}{\lambda}\frac{dF}{dC} \quad \text{(direct disutility of illness)}$$

$$- w^*W\frac{dF}{dC} \quad \begin{array}{l}\text{(lost work productivity,} \\ \text{evaluated at the} \\ \text{before-tax wage rate)}\end{array}$$

$- Fw^*W_C$ (lost work time, evaluated at the before-tax wage rate)

$- FwL_C$ (the value of lost leisure, evaluated at the after-tax wage)

$+ m\dfrac{dS}{dC}$ (medical expenses)

$+ D_C$ (defensive expenditures)   (3-11)

One noteworthy aspect of this result is that there is no lost "leisure productivity" term corresponding to the lost "work productivity" term w^*WdF/dC. In this model, the individual, by assumption, receives less satisfaction when illness strikes, but this effect is captured in the first term, the direct disutility of illness.

The procedure for applying this model to a contamination episode is as follows. The most important events in estimating losses to individuals are the points at which the water supply becomes contaminated, the contamination becomes known to the public, and the contamination ends. These three events define two intervals during which individuals have different information, and hence behave differently. These differences, in turn, affect the nature of the damages incurred.

In the first interval, individuals incur no incremental avoidance costs because they are unaware of anything to avoid. Therefore $D_C = 0$ in equation (3-11) above, and the estimate of losses is found by evaluating the other terms.[2]

In the second interval, individuals can avoid illness by taking averting action. Thus, the losses associated with reduced productivity or incapacity owing to acute illness are less important, and defensive expenditures correspondingly more important. Indeed, in the second interval, where near-perfect protection is possible, the only term of equation (3-11) assumed to be nonzero is D_C. Unfortunately, D_C is an inadequate guide for estimating losses in the second interval for two reasons.

---

[2]Caution is in order, however, because error may be introduced if these marginal conditions are used to evaluate a nonmarginal change. For example, inframarginal hours of leisure are in all likelihood more highly valued than those at the margin. For life-threatening or chronic illnesses requiring a lengthy convalescence or major change in lifestyle, this underestimate is likely to be a major source of error. However, for acute nonlethal infections, such as giardiasis, this effect is probably of minor importance.

First, the model suggests that a substance D can be bought that protects against illness. In fact, people respond to water contamination by changing their consumption pattern—mainly by substituting clean water for contaminated water. In this context, averting behavior consists of either securing water from an uncontaminated source or treating the contaminated tap water by filtering or boiling it. Because consumption of some goods (bottled water, for example) changes from zero to some positive amount, the assumption of an interior maximum made in the preceding model is no longer justified. Second and more important, actions taken to avoid drinking contaminated water represent such a major change in household activities that extrapolation of willingness to pay from the marginal conditions does not seem justified.

The next section describes a model from which the economic damages of contamination can be determined.

VALUATION OF AVOIDANCE COSTS

In a paper on the valuation of avoidance costs, Courant and Porter (1981) asked whether observed defensive expenditures made by an individual to avoid exposure to an increase in pollution were a good estimate of the individual's willingness to pay to avoid exposure. Their conclusion was essentially negative. They found that not only were defensive expenditures likely to be poor estimates of willingness to pay, but that in many cases one could not say whether such expenditures were an overestimate or an underestimate.

By making a couple of assumptions that do not seem unreasonable for the problem of drinking water contamination, an expression that gives observable upper and lower bounds for the willingness to pay to avoid contaminated water can be derived. Suppose an individual's utility function can be represented by

$$U = U[x, W, B, S(W, C)], \qquad (3\text{-}12)$$

where

x is a composite of all goods except water,

W is the quantity of tap water consumed,

B is the quantity of bottled water consumed (or water from a source other than the tap, including boiling),

S is the individual's expected illness caused by exposure to the contaminating substance in the tap water,

C is the concentration of the contaminating substance in the water supply.

The following assumptions can be made about an individual's utility for goods and services:

$$\frac{\partial U}{\partial x} > 0, \qquad \frac{\partial^2 U}{\partial x^2} < 0$$

$$\frac{\partial U}{\partial W} > 0, \qquad \frac{\partial^2 U}{\partial W^2} < 0$$

$$\frac{\partial U}{\partial B} > 0, \qquad \frac{\partial^2 U}{\partial B^2} < 0 \qquad\qquad (3\text{-}13)$$

$$\frac{\partial U}{\partial S} > 0, \qquad \frac{\partial^2 U}{\partial S^2} < 0$$

$$\frac{\partial S}{\partial W} \geq 0, \qquad \frac{\partial S}{\partial C} \geq 0$$

The willingness to pay to avoid contaminated water is the money necessary to restore the individual to his initial utility level before contamination. Let p_x, p_W, and p_B, respectively, denote the prices of x, W, and B, and assume $p_W < p_B$, that is, that tap water is less expensive than bottled water.[3] The individual's expenditure-minimizing problem (the dual of the utility maximizing problem)[4] is to minimize

$$p_x x + p_W W + p_B B, \qquad\qquad (3\text{-}14)$$

subject to the constraint

$$U[x, W, B, S(W, C)] \geq U_0 \qquad\qquad (3\text{-}15)$$

Forming the Lagrangian,

$$\pounds = p_x x + p_W W + p_B B + \lambda(U_0 - U), \qquad\qquad (3\text{-}16)$$

---

[3]The price of bottled water includes not only what is paid at the store, but the cost of getting it home plus whatever inconvenience there may be from having the water come out of a bottle rather than a tap. Note also that because the price p_B of the alternative source is fixed, this model does not apply to solutions requiring investment, such as filtration systems.

[4]Solving the expenditure-minimizing problem is equivalent to solving the utility-maximizing problem, as in the previous section, and is slightly more convenient in the current case.

we have the following first order conditions:

$$\frac{\partial \pounds}{\partial x} = p_x - \lambda \frac{\partial U}{\partial x} = 0 \tag{3-16a}$$

$$\frac{\partial \pounds}{\partial W} = p_W - \lambda \frac{\partial U}{\partial W} - \lambda \frac{\partial U}{\partial S} \frac{\partial S}{\partial W} = 0 \tag{3-16b}$$

$$\frac{\partial \pounds}{\partial B} = p_B - \lambda \frac{\partial U}{\partial B} \geq 0, \tag{3-16c}$$

with

$$B\left(p_B - \lambda \frac{\partial U}{\partial B}\right) = 0$$

$$\frac{\partial \pounds}{\partial \lambda} = U_0 - U = 0$$
$$\text{and } \lambda > 0 \tag{3-16d}$$

Recognizing that bottled water consumption B may be zero, equation (3-16c) is the Kuhn–Tucker condition rather than the classical condition.[5] Now form the expenditure function $E(p_x, p_W, p_B, C, U_0)$. The marginal willingness to pay to avoid contamination is given by $\partial E/\partial C$.

Because not all the first order conditions in equation (3-16) are zero, the envelope theorem cannot be applied directly. However, $\partial E/\partial C$ can be evaluated directly, yielding

$$\frac{\partial E}{\partial C} = \frac{\partial \pounds}{\partial B} \frac{\partial B}{\partial C} + \frac{\partial \pounds}{\partial C} = s \frac{\partial B}{\partial C} - \lambda \frac{\partial U}{\partial S} \frac{\partial S}{\partial C} \tag{3-17}$$

where s is the slack variable,

$$s = p_B - \lambda \frac{\partial U}{\partial B} \geq 0 \tag{3-16c$'$}$$

We claim, nonetheless, that $s \frac{\partial B}{\partial C} = 0$. From equation (3-16c$'$), either $s = 0$ or $s > 0$. If $s = 0$, then there is nothing to prove. But suppose

---

[5]The Kuhn–Tucker conditions allow for the possibility of a corner solution rather than the interior solution required by the classical conditions. See chapter 4 of Intriligator (1971).

$s > 0$. With all parameters except C fixed, s is a function of C. Therefore, $s^* = s(C^*) > 0$ can be assumed for some contamination level C^*. There is an open neighborhood of s^*, not containing 0,

$$N_\varepsilon(s^*) = \{s \,|\, 0 < s^* - \varepsilon < s < s^* + \varepsilon\}$$

Corresponding to this neighborhood is a neighborhood of C^*,

$$N_\delta(C^*) = \{C \,|\, C^* - \delta < C < C^* + \delta\},$$

so that whenever $C\varepsilon N_\delta(C^*)$, $s\varepsilon N_\varepsilon(S^*)$. But by the Kuhn–Tucker conditions, we know that if $s > 0$, $B = 0$. This means that $B = 0$ for every $C_\varepsilon N_\delta(C^*)$. But if $B = 0$ in an interval surrounding C^*,

$$\frac{\partial B}{\partial C}(C^*) = 0$$

In other words, either $s = 0$ or $\dfrac{\partial B}{\partial C} = 0$.

Thus, equation (3-17) can be rewritten as

$$\frac{\partial E}{\partial C} = -\lambda \frac{\partial U}{\partial S} \frac{\partial S}{\partial C} \tag{3-17'}$$

This expression cannot be evaluated because it contains terms (λ and $\partial U/\partial S$) that cannot be observed. However, if two additional assumptions are made, the marginal willingness to pay can be bracketed between two readily observable expressions.

The first assumption is that individuals adjust their water consumption to avoid any additional expected illness. (This assumption may not always be valid but appears to be almost always true for *Giardia* contamination.) If so, then the total derivative of S with respect to C, dS/dC, is zero. But

$$0 = \frac{dS}{dC} = \frac{\partial S}{\partial W} \frac{\partial W}{\partial C} + \frac{\partial S}{\partial C} \quad \text{or} \quad \frac{\partial S}{\partial C} = -\frac{\partial S}{\partial W} \frac{\partial W}{\partial C} \tag{3-18}$$

Substituting into equation (3-17'), we have

$$\frac{\partial E}{\partial C} = -\lambda \frac{\partial U}{\partial S} \frac{\partial S}{\partial C} = \lambda \frac{\partial U}{\partial S} \frac{\partial S}{\partial W} \frac{\partial W}{\partial C}$$

But from equation (3-16b),

$$\lambda \frac{\partial U}{\partial S} \frac{\partial S}{\partial W} = p_W - \lambda \frac{\partial U}{\partial W}$$

Hence

$$\frac{\partial E}{\partial C} = \left(p_W - \lambda \frac{\partial U}{\partial W} \right) \frac{\partial W}{\partial C} \qquad (3\text{-}19)$$

The second assumption is that tap water and bottled water provide, at the margin, the same utility,[6] so that

$$\frac{\partial U}{\partial B} = \frac{\partial U}{\partial W}$$

Then we can write

$$\frac{\partial E}{\partial C} = \left(p_W - \lambda \frac{\partial U}{\partial B} \right) \frac{\partial W}{\partial C}$$

But from equation (3-16c'), $\lambda \frac{\partial U}{\partial B} = p_B - s$, so that

$$\frac{\partial E}{\partial C} = (p_B - s - p_W) \left(-\frac{\partial W}{\partial C} \right)$$

When $s = 0$, the marginal willingness to pay to avoid contamination is the difference in price between bottled and tap water multiplied by the marginal reduction in tap water consumption. When $s > 0$, the expression $(p_B - s)$ can be thought of as the price bottled water would have to be before the individual would begin to consume it (at the given contamination level). The marginal willingness to pay to avoid contamination is therefore the willingness to pay for bottled water, less the price of tap water, times the marginal change in tap water consumption.

Now, suppose contamination increases from C_0 to C_1. Assume $s(C_0) > 0$ and $s(C_1) = 0$ (that is, somewhere in the interval (C_0, C_1) the purchase of bottled water begins).

Let

---

[6]We do not know, a priori, whether bottled water has a greater marginal utility than tap water, and there are reasons for thinking either might be true. On the one hand, bottled water might be more palatable than tap water; on the other, tap water might be more convenient because it is under pressure.

$$\overline{C} = glb\{C \,|\, s(C) = 0\},$$

where *glb* signifies greatest lower bound. The total willingness to pay (*WTP*) to avoid the increase in pollution is

$$WTP = \int_{C_0}^{C_1} \frac{\partial E}{\partial C} \, dC$$

$$= \int_{C_0}^{\overline{C}} (p_B - s - p_W)\left(-\frac{\partial W}{\partial C}\right) dC$$

$$+ \int_{\overline{C}}^{C_1} (p_B - p_W)\left(-\frac{\partial W}{\partial C}\right) dC \qquad (3\text{-}20)$$

To get observable bounds on willingness to pay, we note that since $\partial U/\partial B = \partial U/\partial W$,

$$s = p_B - \lambda\frac{\partial U}{\partial B} = p_B - \lambda\frac{\partial U}{\partial W}$$

$$= p_B - p_W + \lambda\frac{\partial U}{\partial S}\frac{\partial S}{\partial W} \quad \text{(from 3-16b)} \qquad (3\text{-}21)$$

But $\lambda > 0$, $\partial U/\partial S < 0$, $\partial S/\partial W \geq 0$ by assumption (3-13), so that the last term in equation (3-21) is negative. Hence

$$0 \leq s \leq p_B - p_W.$$

Therefore, the first integral on the right-hand side of equation (3-20) is nonnegative, so that

$$WTP \geq \int_{\overline{C}}^{C_1} (p_B - p_W)\left(-\frac{\partial W}{\partial C}\right) dC \qquad (3\text{-}22)$$

Also,

$$p_B - s - p_W \leq p_B - p_W$$

so that

$$WTP \leq \int_{C_0}^{\overline{C}} (p_B - p_W)\left(-\frac{\partial W}{\partial C}\right) dC + \int_{\overline{C}}^{C_1} (p_B - p_W)\left(-\frac{\partial W}{\partial C}\right) dC$$

$$\leq \int_{C_0}^{C_1} (p_B - p_W)\left(-\frac{\partial W}{\partial C}\right) dC \qquad (3\text{-}23)$$

Equations (3-22) and (3-23) can be integrated and rewritten as

$$(p_B - p_W)(W(C_0) - W(C_1)) \geq WTP$$

$$\geq (p_B - p_W)(W(\overline{C}) - W(C_1)) \qquad (3\text{-}24)$$

We cannot directly evaluate the rightmost term of equation (3-24) because we cannot observe $\overline{C}$, the point at which bottled water consumption would begin if individuals were compensated for the additional costs of obtaining drinking water. However, since the price of bottled water exceeds the price of tap water, the decline in tap water consumed beyond $W(\overline{C})$ can be assumed to be at least as large as the increase in bottled water consumed; that is,

$$W(\overline{C}) - W(C_1) \geq B(C_1) - B(\overline{C}) = B(C_1), \qquad (3\text{-}25)$$

since $B(\overline{C}) = 0$. Hence

$$(p_B - p_W)(W(C_0) - W(C_1)) \geq WTP \geq (p_B - p_W)B(C_1) \qquad (3\text{-}26)$$

Equation (3-26) shows that willingness to pay to avoid a change in contamination can be bracketed by the difference in price times the decrement in tap water consumption and the difference in price times the increment in bottled water consumption.

The argument is illustrated in figure 3-1. The bottom quadrant on this diagram shows the relationship between the concentration of contaminant C and tap water consumption Z. Initially it is assumed that bottled water consumption is zero; that is, $B(C_0) = 0$. As the level of contamination increases, Z decreases as individuals act to avoid contamination (or to keep contamination constant). This decrease in Z is inconvenient, and people would therefore be willing to pay a price greater than the normal price of tap water, p_Z, in order to obtain additional safe water. The upper quadrant gives the relationship between tap water consumption Z and this price premium. As the contamination level increases, the price premium increases until it reaches $p_B - p_Z$, the price difference between bottled water and tap water. That occurs at the contamination level $\overline{C}$, and water consumption $\overline{Z}$. With further increases in contamination, tap water consumption continues to fall. The premium does not increase, however, for the household does not have to pay more than p_B to purchase bottled water. With contamination at C_1, tap water consumption falls to Z_1. As shown in the figure, the bottled water consumption B may be less than the decrement in tap water consumption, $\overline{Z} - Z_1$.

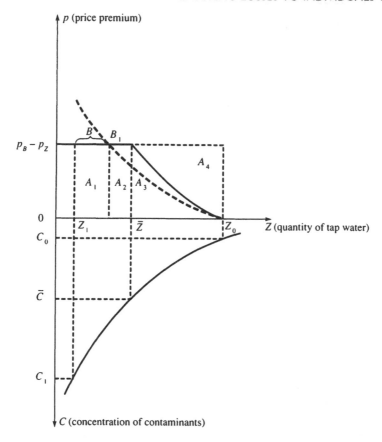

Figure 3-1. Willingness to pay to avoid contaminated water.

The "true" willingness to pay to avoid a change from C_0 to C_1 is the area under the price premium curve in the upper quadrant, or the areas marked A_1, A_2, and A_3. The lower and upper bounds mentioned above are, respectively, A_1, and $A_1 + A_2 + A_3 + A_4$.[7] For a linear demand curve, the willingness to pay is found by averaging the upper and lower bounds. (In the empirical work in chapter 6, this average is referred to as the "best estimate.")

It is interesting to compare these bounds on willingness to pay with defensive expenditures. Defensive expenditures are defined to be equal

---

[7]The premium curve is, in effect, the (Hicksian) demand curve for clean water. The Marshallian demand curve passes through the points Z_0 and B_1, as indicated by the dotted line in the figure.

to the change in expenditures brought on by the action, which is expressed as

$$D = p_W(\Delta W) + p_B(\Delta B),$$

where ΔW and ΔB are the changes in tap water and bottled water consumption, respectively. We can write

$$\Delta W + \Delta B + u = 0$$

for some $u > 0$. Hence, from equation (3-26),

$$WTP \geq (p_B - p_W)\Delta B = p_B\Delta B - p_W(-\Delta W - u)$$
$$= p_B\Delta B + p_W\Delta W + p_W u \geq D$$

Therefore, our results provide a closer estimate of the true willingness to pay than that provided by measuring only defensive expenditures.

EMPIRICAL CONSIDERATIONS

The models presented above yield the following results. First, losses for those who fall ill due to the outbreak are the sum of five elements: the value of work and leisure time lost due to illness, medical costs, defensive expenditures (measures taken to reduce future exposure to contaminants), the value of pain and suffering, and the value of special leisure activities that cannot be rescheduled. Second, losses for those who do not become ill but who nonetheless take measures to avoid contamination are bracketed by the difference in price between bottled and tap water multiplied both by the increment of bottled water consumption and by the decrement in tap water consumption.

The behavioral theory developed in the previous two sections, in particular, equations (3-11), (3-22), and (3-24), provides the framework for estimating an individual's willingness to pay to avoid an outbreak of waterborne disease. As a general proposition, almost no data were available before this study to capture the benefits of avoiding an outbreak. Economic data were lacking because the state health departments and the Centers for Disease Control in Atlanta, which are responsible for nearly all data collection, are concerned primarily with identifying the nature of the illness, its proximate cause, and the factors that lead to an increased likelihood of illness. This epidemiological perspective is quite different from an economic perspective, where the interest is in valuing the benefits of avoiding an outbreak.

To obtain much of the information needed for benefit measurement, we distributed a questionnaire on morbidity to people with confirmed cases of giardiasis in the outbreak area and a questionnaire on averting behavior to a random sample of people in the area. Those questionnaires are reproduced in appendixes 6-A and 6-B. Chapter 6 describes and analyzes the responses.

Losses Due to Illness

Work Lost   As noted above, a value is attached to lost productivity by assuming that the marginal product (the value of a day's output con- tributed by a worker) is the before-tax wage rate. Then, lost workdays are multiplied by the individual's (self-reported) wage rate to obtain losses to the individual and society at large associated with lost work. A similar valuation procedure is found in most benefit studies of pol- lution control, for example, in Ostro (1983). It is important to note, however, that a sick worker's output can be affected by disease even if the worker is not quite sick enough to stay at home. That effect can be particularly significant for some common waterborne diseases such as giardiasis or salmonellosis, which can have an intermittent or chronic stage lasting several weeks. We attempted to learn more about this economic effect by asking sick individuals to rate their work performance on days when they were ill but nonetheless went to work.

People who work in nonmarket activities, primarily homemakers, earn implicit wages, which should also be counted as losses. Although the questionnaire does not elicit this information directly, the time lost from nonmarket productive activities can be inferred from information on the length and severity of sickness of people who engage in such activities and from information on the proportion of work time lost to total time sick for those involved in market activities. The value of nonmarket work time lost is derived from estimates of the value of household work. Two alternative estimation approaches were consid- ered. First, the value of a homemaker's time is the opportunity cost (to the homemaker) of not working outside the home, as approximated by the after-tax wage rate at a job for which the homemaker is qualified. In this study, that rate is taken to be the average after-tax wage rate for the study area. Alternatively, the value of household work can be estimated by what it would cost to obtain comparable services (such as housecleaning and baby-sitting) in the marketplace. Such estimates have been produced, for example, by Cooper and Rice (1976), Murphy (1982), and Peskin (1983).

We also consider the work time losses of well family members who care for sick family members. Sickness in children and the elderly is quite likely to necessitate expenses of time on the part of others, say, to take them to the doctor or to stay home to care for them (see Salkever, 1980). Sickness in an adult may also cause other family members to miss work (as well as leisure). These losses are identified by the questionnaire.

Time spent in school is also a productive activity, in that investment is being made in human capital, an investment that will bear returns in the form of higher wages later in life. Although data on school time lost because of giardiasis were gathered, an attempt to value such losses was not made. Because a bout of giardiasis is relatively short-lived, the rate of human capital accumulation is unlikely to be appreciably affected.

Medical Costs   To measure medical costs involves obtaining knowledge of the type, price, and quantity of medications prescribed, the quantity and price of laboratory samples, the number and price of visits to the doctor, hospital, and other medical professionals, and the distance and time taken to seek such care. Costs associated with misdiagnoses and side effects of medication (treatment) might also be included here. All this information, except that on prices, is taken from the survey. Price data were collected from pharmacies, doctors, clinics, and hospitals in the outbreak area. Assumptions based on information provided in the questionnaire were made about a typical course of medication and the type of services that would be received during a typical doctor or hospital visit for diagnosis and treatment of giardiasis.

The valuation of costs indirectly associated with doctor or hospital visits, such as those incurred traveling to and waiting in the office, requires some discussion. The survey data on distance traveled can be multiplied by a per mile charge to value transportation costs. To use the survey data on travel time requires a value to be set on an individual's time. While the wage rate is often used, it does not apply to nonworkers; a wage rate corresponding to the opportunity cost of their time is used here to value the time of ill nonworkers. The value workers place on an hour of leisure and an hour of work are not likely to be identical at the margin. Workdays requiring a set number of hours constrain the individual's choice of work time (Harrington and Portney, 1984) and thus break the equivalence between the marginal value of leisure and work time. For some individuals, the marginal value of leisure may exceed that of work time; for others, it may be less.

Forgone Leisure   The last term in equation (3-11) calls for measurements of the incremental value of lost leisure for each activity where the marginal value exceeds the wage rate. The model implicitly values other leisure time at the wage rate. Data limitations preclude valuation of the highly valued leisure activities. In the absence of this information, all lost leisure time is valued at the after-tax wage rate or at a rate adjusted for productivity loss (see chapter 6), a procedure that may undervalue certain leisure activities and overvalue others. The after-tax wage rate is used because it represents, at the margin, what one must give up in income to expand leisure activities, assuming work and leisure can be traded off continuously.

Losses Due to Averting Actions—The Cost of Alternative Water Supplies

For those who do not become ill, the major losses of an outbreak of a waterborne disease are probably the extra cost and inconvenience of obtaining water from an uncontaminated source. From equation (3-25), the information necessary to calculate willingness to pay to avoid the contaminated water includes the normal price of tap water and quantity used before the outbreak and the price and quantity of water obtained from the alternative source. The results of a telephone survey of fifty households in the affected area in Luzerne County showed that individuals and households boiled tap water, purchased bottled water, obtained uncontaminated water free of charge from distribution points established by the communities and the water utility, or combined these options. According to government officials in the affected area, virtually no one invested in home filtration systems.

The price p_B of bottled water includes not only the price paid at the store, but the time and transportation costs, at the margin, of bringing the water home. Inasmuch as people may buy bottled water at the supermarket as part of their regular shopping routine, this marginal travel cost may be very low. We must therefore consider whether the trips made to obtain bottled water would have been made anyway and whether the bottled water was purchased as a result of the outbreak.

The same considerations apply for water obtained from distribution points, except that the purchase price is zero[8] and the marginal transportation cost is positive since the distribution points were at public buildings, such as firehouses, where individuals do not shop.

The main costs of boiled water are the energy costs and the time required to prepare it. As shown in the appendix to this chapter, the average energy and time costs are about five cents and twelve minutes a gallon, respectively, the latter of which can be valued using estimates

---

[8]Provision of such water supplies is a cost, of course, but not to the individual. This cost will be accounted for as a community cost, as discussed in chapter 4.

from the literature for the value of household work. Again, however, it is possible to combine boiling water with other tasks, such as meal preparation, so that the incremental time required may be virtually zero.

REFERENCES

Baker, Edward L., Wendy Patterson, Stephen Van Allmen, and Jane Fleming. 1979. "Economic Impact of a Community-Wide Waterborne Outbreak of Gastrointestinal Illness," *Public Health Briefs* vol. 69, no. 5, pp. 501–502.

Becker, G. S. 1965. "A Theory of the Allocation of Time." *The Economic Journal* vol. 75, pp. 493–517.

Cooper, B. S., and D. P. Rice. 1976. "The Economic Cost of Illness Revisited," *Social Security Bulletin* February, pp. 21–36.

Courant, Paul, and Richard Porter. 1981. "Averting Expenditures and the Cost of Pollution," *Journal of Environmental Economics and Management* vol. 8, no. 4, pp. 321–329.

Harrington, Winston, and Paul Portney. 1987. "Valuing the Benefits of Health and Safety Regulation," *Journal of Urban Economics* vol. 22, pp. 101–112.

Intriligator, Michael D. 1971. *Mathematical Optimization and Economic Theory* (Englewood Cliffs, N.J., Prentice Hall).

Levy, Barry S., and Ward McIntire. 1974. "The Economic Impact of a Food-Borne Salmonellosis Outbreak," *Journal of the American Medical Association* vol. 230, no. 9, pp. 1281–1282.

Murphy, Martin. 1982. "The Value of Household Work in the United States, 1976," in *Measuring the Nonmarket Economic Activity: BEA Working Papers* (Washington, D.C., Bureau of Economic Analysis, U.S. Department of Commerce).

Ostro, Bart. 1983. "The Effects of Air Pollution on Work Loss and Morbidity," *Journal of Environmental Economics and Management* vol. 10, no. 4., pp. 371–382.

Peskin, Janice. 1983. "The Value of Household Work in the 1980s." Paper presented at the American Statistical Association Meetings, Toronto, Canada, August 15–18, 1983.

Russell, Clifford S., David Arey, and Robert Kates. 1970. *Drought and Water Supply: Implications of the Massachusetts Experience for Municipal Planning* (Baltimore, Md., Johns Hopkins University Press for Resources for the Future).

Salkever, David. 1980. "Effects of Children's Health on Maternal Hours of Work: A Preliminary Analysis," *Southern Economic Journal* vol. 47, no. 1, pp. 156–166.

Schwab, Paul M. 1968. "Economic Cost of St. Louis Encephalitis Epidemic in Dallas, Texas, 1966," *Public Health Reports* vol. 83, no. 10, pp. 860–866.

Winkler, Donald R. 1980. "The Effects of Sick Leave Policy on Teacher Absenteeism," *Industrial and Labor Relations Review* vol. 33, no. 2, pp. 232–240.

Appendix 3-A

Cost of Boiling Water

The cost of boiling water for a household, factory, or commercial establishment during an outbreak of giardiasis is given by the following formula:

$$C = [p_E\, E(V) + kp_T\, t(V)]\, nD, \qquad (3\text{-}A\text{-}1)$$

where

p_E = the price of energy (\$/kilowatt-hour[kWh]),

$E(V)$ = the amount of energy (kWh) required to boil V gallons of water for three minutes,

V = the batch size, or the volume (gallons) of water to be boiled at any one time,

k = a factor ($0 \le k \le 1$), to be discussed below,

p_T = the value of time (\$/hr),

$t(V)$ = the amount of time required to boil V gallons (hrs.),

n = the number of batches prepared per day,

D = the duration of the outbreak (days).

The first term in parentheses represents the energy cost per day; the second term, the time cost per day.

ENERGY COST (PER GALLON)

Assuming no loss of energy, the energy (in kWh) required to raise V gallons of water to the boiling temperature (212°F) is given by

$$E = 8.34\,\frac{\text{lbs}}{\text{gal}} \times V\text{ gal} \times \frac{1\text{ BTU}}{\text{lb-°F}} \times (212 - T_{IN})$$

$$\times \frac{1\text{ kWh}}{3,412\text{ BTU}}\, 0.0024\, V\, (212 - T_{IN})$$

where T_{IN} is the temperature of the intake water. For example, if $T_{IN} = 80°F$, then

$$E_{IN} = 0.323\text{ kWh/gal}.$$

However, energy losses through thermal conductivity to the surrounding air and evaporation are quite significant. In an experiment conducted in the kitchen of one of the authors, the time required to boil water was found to depend on the volume, the temperature of the tap water, the power output of the burner, the geometry of the pan, and whether the pan was covered. Modest economies of scale with respect to batch size V were also apparent. For example, for an uncovered ten-inch pan, a 2,600-watt burner, and a beginning water temperature of 80°F, a linear relationship between the time to boiling (seconds) and the volume of water (gallons) was found:

$$t(V) = 76.3 + 688V, \text{ (for } T_{IN} = 80°F) \tag{3-A-2}$$

Given the burner rating, this translates into the following relationship between the energy required (in kWh) and the volume of water (in gallons):

$$E(V) = 0.055 + 0.5V, \tag{3-A-3}$$

Thus, the energy required to raise the temperature of half a gallon of water ($V = 0.5$ gal) at 80°F to 212°F is

$$E(V) = 0.055 + 0.5(0.5)$$

$$E(V) = 0.305 \text{ kWh.}$$

The energy efficiency of boiling half a gallon of water in a ten-inch, uncovered pan on a 2,600-watt burner can be calculated by dividing the energy delivered to the water in the pan, E_{IN}, by the total energy consumed, $E(V)$. Since E_{IN} equals 0.161 kWh and $E(V)$ equals 0.305 kWh, the efficiency is 53 percent.

The per unit energy cost of boiling water depends on the price of the input energy, as indicated in equation (3-A-1). The total cost of raising V gallons of water to 212°F may be expressed as

$$C_E = p_E E(V) = p_E (0.055 + 0.5V) \tag{3-A-4}$$

where p_E is the cost of energy in dollars per kWh, $E(V)$ is the energy required in kWh, and V is the volume in gallons. For example, if electricity costs 7 cents a kWh, the energy cost of raising 0.5 gallons of water from 80°F to 212°F is 2.14 cents, or 4.3 cents a gallon.

It should be noted that equation (3-A-4) is valid only for the empirical conditions used in the experiment—a 2,600 watt burner, an uncovered, ten-inch pan, and a beginning water temperature of 80°F. Moreover, it provides the energy cost to reach the boiling point, not the cost to boil water for one minute, the length of time recommended by public health officials to kill *Giardia* cysts. This example is intended to be illustrative.

TIME COST

Recall equation (3-A-2), which is a relationship between volume of water (in gallons) and the time (in seconds) required for a 2,600-watt burner to raise the temperature of water from 80°F to 212°F. Translated into hours, the expression becomes

$$t(V) = 0.021 + 0.19V \tag{3-A-5}$$

To complete the calculation of time cost, an estimate of the value of time is needed. This calculation is plagued not only by the usual difficulties of estimating the value of leisure time, but also by the fact that boiling water can be carried on simultaneously with other activities, such as preparing meals or reading the newspaper. *All* the value of the time required for boiling cannot therefore be attributed to the cost of boiling water. The time cost of boiling water can therefore be written as

$$C_T = kP_T t(V) \tag{3-A-6}$$

where $0 \leq k \leq 1$ and p_T is the value of leisure time.

The time cost can be bracketed by assuming $k = 0$ (that is, the marginal cost is zero) and $k = 1$. A tighter upper bound may be obtained by considering the cost of obtaining potable water from another source. For example, if the full price of a gallon of bottled water is x (retail price plus transportation and time cost), it is reasonable to suppose that the full price of boiled water is no more than x, assuming that boiled water is no more desirable than bottled water.

TOTAL COST

Thus, the cost of boiling water can be calculated by adding the cost of energy to the cost of time. Although the energy cost can be estimated relatively easily through empirical studies, the time cost is more difficult to estimate because the task of boiling water can be performed simultaneously with other tasks or leisure activities. In an attempt to take this into account, the variable constant, k, has been included in the equation of time cost. A lower bound on the time cost is obtained when $k = 0$. An upper bound is obtained when $k = 1$.

4 / Measuring Losses to Businesses and Government Agencies

Although the losses suffered by individuals and the steps taken by public authorities to curtail an outbreak of giardiasis probably attract most of the media attention, businesses, school districts, government agencies, and the water supply utility in an affected area also experience losses. This chapter develops methods for measuring these losses to society based on the difference between the sum of the producer and consumer surpluses before and after an outbreak.[1]

LOSSES TO BUSINESSES

This section concerns the measurement of losses to businesses in an affected area. The theory is presented first, followed by discussions of the application of theory, problems of empirical estimation, and complications.

---

[1]This analysis measures actual losses from an outbreak, rather than estimating losses from a possible future outbreak. The former is an ex post analysis, whereas the latter is ex ante. The appropriate reference against which to measure losses is the conditions that would have existed in this community had no outbreak occurred. This analysis compares consumer and producer surpluses with and without the outbreak and uses the differences in the surpluses as a measure of the losses. Chapter 8 uses alternative probabilities to convert the ex post measures of losses to ex ante measures of losses.

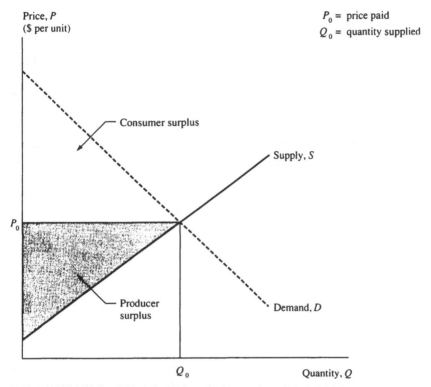

Figure 4-1. **Typical supply and demand functions for goods and services showing consumer and producer surplus.**

Theory

Principles of welfare economics can be used to estimate the social losses associated with changes in the supply of and demand for goods and services provided by businesses during an outbreak of giardiasis. These principles define social losses as changes in the sum of consumer and producer surpluses in the affected markets.[2] *Consumer surplus* is defined as the amount consumers would be willing to pay for goods and services minus the amount they actually paid for them. Consumer surplus is represented in figure 4-1 by the area between the demand function, D, and the price paid, P_0. *Producer surplus* is defined as the amount producers actually receive for their goods and services minus the costs of

---

[2]The correct measures of social welfare change are the aggregate willingness to pay to avoid the outbreak and the aggregate willingness to accept compensation for the outbreak. These concepts are termed the equivalent and the compensating variation. The sum of the consumer and producer surpluses is likely to be a close approximation of these theoretically correct conceptual measures (Willig, 1976).

production. Producer surplus is represented in figure 4-1 by the area between the supply function, S, and the price paid, P_0.

Reductions in the demand for goods and services can occur when illness, or the fear of becoming ill, drives consumers away from businesses. Restaurants and other businesses providing food and drink are particularly vulnerable to this effect. Service losses may be more difficult to recoup after an outbreak has ended than are losses associated with goods production. Restaurants, for example, cannot store meals for future sale and, unlike local goods-producing businesses which can sell as much as they can produce at the regional or national price, restaurants, as a group, have a limited market.

Reductions in producer surplus can be caused by productivity losses when workers are too ill to work as productively as usual or to work at all. Businesses such as food producers that use water in production may also incur additional costs to treat the contaminated water or to obtain an alternative source of uncontaminated water.

Increases in the demand for goods and services resulting from the outbreak cannot be used to offset social losses. This statement may seem counterintuitive at first. However, the "positive" effects of an outbreak on individual businesses, such as increased sales of bottled water and health care services, are not benefits at all, but expenditures made to neutralize the effects of the outbreak. More important, these expenditures are made possible by giving up other goods and services that consumers would have preferred to buy in the absence of the outbreak.[3]

In measuring losses to businesses in a typical outbreak area, two types of market effects are distinguished: a change in the demand for goods and services (a shift in the demand curve), and a change in the supply of goods and services (a shift in the supply curve). In this theoretical development, let us consider the two effects separately first, and then jointly. A simultaneous shift in supply and demand appears to capture best the market conditions for the restaurants and bars in the affected area. These establishments appear to have borne the brunt of the losses to businesses.

Change in Demand Assuming in all cases a downward-sloping demand function, individual cases can be differentiated by the slope of the supply curve and by whether prices and quantities traded adjust when demand changes (that is, when the demand curve shifts). Three possibilities are most likely. Case 1 involves an upward-sloping supply

---

[3]Confusion may arise here because measures of expenditures on goods and services, such as gross national product (GNP), are sometimes viewed, incorrectly, as measures of social welfare. Attempts have been made to purge the purchase of goods used to neutralize social "bads," such as burglar alarms, health care (except prevention), and pollution control equipment, from the GNP accounts to allow this measure to reflect social welfare more closely.

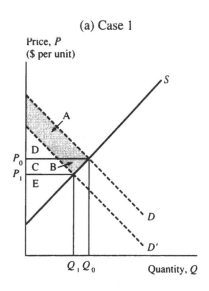

(a) Case 1

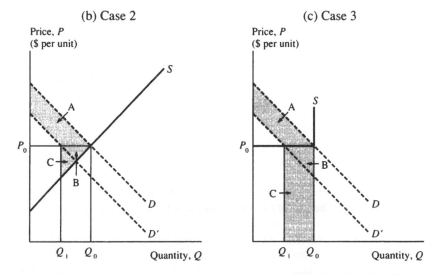

Figure 4-2. Losses in consumer and producer surplus due to a shift in demand. (a) Case 1: upward-sloping supply function with price change. (b) Case 2: upward-sloping supply function with no price change. (c) Case 3: L-shaped supply function with no price change.

curve. Given the downward-sloping demand curve (D) and the upward-sloping supply curve (S) in figure 4-2 (a), if demand shifts (to D') and prices adjust to the change in demand (from P_0 to P_1), the social loss involves a change in both consumer surplus and producer surplus, depicted by the shaded area A + B. Note that consumer surplus, which was initially A + D, is now D + C. The change in surplus is A − C. Producer surplus, which originally was B + C + E, is now E. The change is B + C. Thus, consumers lose the area A − C and producers lose the area C + B. From a social perspective, the loss of C by producers is offset by the gain of C by consumers. The net social loss, therefore, is represented by the area A + B.

If changes in demand are small, short-lived, and localized, prices may not be affected. In this second case, the social loss again involves changes in both consumer and producer surplus, but the net effect is larger than before. This loss is depicted by the shaded area A + B + C in figure 4-2(b).

In both case 1 and case 2, knowledge of the slopes of the demand and supply functions (sometimes specified by demand and supply elasticities), and the pre-outbreak and post-outbreak prices and quantities of goods and services are needed to calculate the losses.

The important lesson to learn from these two theoretical developments is that the loss in revenues, ($P_0Q_0 − P_1Q_1$), is not an appropriate measure of the social losses associated with the effects of an outbreak on businesses. Loss of revenue is one measure of losses to businesses, but it bears little relationship to actual social welfare losses, which may be larger or smaller than the revenue losses.

Case 3, depicted in figure 4-2(c), involves a supply curve that is horizontal up to a capacity constraint and vertical at that constraint. This special case applies to businesses where the marginal cost of providing an additional unit of a product between a nonzero lower bound and a maximum capacity upper bound is zero or close to it. An example is a movie theater. The supply curve is horizontal at zero price up to the seating capacity, and vertical at that capacity. A loss of one customer to the theater leaves the costs of providing that service virtually unaffected.[4] For this special case, short-run changes in demand caused by an outbreak of giardiasis ordinarily will not affect market prices. That is certainly true for the movie theater example. A downward shift in

---

[4]A vertical supply curve implies that a given output will be supplied regardless of the price it commands in the marketplace. Care must be used in interpreting the vertical supply function, however. The supply and demand functions shown in the figures in this chapter are simplifications of actual supply and demand functions and are used to elucidate the effects of small changes in supply and demand at the margin. They do not provide sufficient information to determine if the business is profitable and thus able to remain in business if prices drop. Information on *total* costs and *total* revenues is necessary to determine the feasibility of continuing to operate as prices change. In general, firms in

demand will reduce revenues to the business. This loss in producer surplus is depicted by the shaded areas B + C in figure 4-2(c). A downward shift in demand will also reduce consumer surplus, which is shown by the shaded area A. The total social loss is the sum of losses in consumer and producer surplus, the shaded areas A + B + C in figure 4-2(c).

To reiterate, the important lesson from theory is that losses in revenue bear little relationship to actual losses in social welfare. Revenue losses, however, can sometimes provide a lower-bound estimate of the social losses. In the case of figure 4-2(c) where prices do not change and the supply curve is vertical, revenue losses are always *less* than total social welfare losses.

Change in Supply   The second type of market effect involves a change in the supply of a good or service. For this development, the demand for a good or service is assumed to be unaffected, but the marginal costs of producing any level of output are assumed to increase. This increase can be attributed to employee illness, the cost of treating water, or the cost of supplying water from an alternative source. Three cases are distinguished: a vertical supply curve representing a capacity constraint, a horizontal supply curve, and an upward-sloping supply curve. Three subcases are distinguished for the upward-sloping supply curve: both prices and quantities change, quantities change but prices do not, and neither prices nor quantities change. The five cases and subcases are represented in figure 4-3.

For cases 1 and 2 with the vertical and the horizontal supply functions (figures 4-3(a) and 4-3(b)), the sum of the consumer and producer losses are represented by the shaded areas, A + B. If one assumes that the demand function is linear between Q_1 and Q_0, these shaded areas may be calculated as areas of rectangles and triangles. For vertical supply functions,

$$\text{Welfare loss} = (Q_0 - Q_1) [P_0 + 1/2 (P_1 - P_0)]$$

For horizontal supply functions,

$$\text{Welfare loss} = (P_1 - P_0) [Q_1 + 1/2 (Q_0 - Q_1)]$$

---

sectors with flat supply curves at zero price up to a capacity constraint and vertical supply curves at that constraint operate with high fixed costs and little or no variable costs. In contrast, firms in sectors with horizontal or nearly horizontal supply curves at finite prices have constant or nearly constant marginal costs and may or may not have high fixed costs. The point is this: the supply and demand analysis in this chapter is a marginal cost analysis, not a total cost analysis. A total cost and revenue analysis is necessary to determine if a business is profitable.

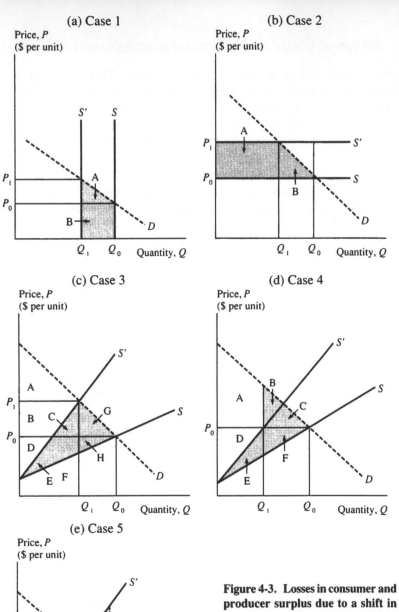

Figure 4-3. Losses in consumer and producer surplus due to a shift in supply. (a) Case 1: vertical supply function. (b) Case 2: horizontal supply function. (c) Case 3: upward-sloping supply function with change in price and quantity. (d) Case 4: upward-sloping supply function with change in quantity but no change in price. (e) Case 5: upward-sloping supply function with no change in price or quantity.

From these expressions, it is evident that both pre- and post-outbreak prices and quantities are needed to calculate losses. However, because price times quantity equals revenue, information on revenues may substitute for direct knowledge of either prices or quantities. Relaxing the assumption of linear demand means that the shape of the demand function between Q_1 and Q_0 must also be known.

The case of the upward-sloping supply function shown in figure 4-3(c) is more complex. When the costs of production increase, prices rise accordingly, and consumers lose a surplus represented by the areas B + C + G in figure 4-3(c). Producers, who before the outbreak were earning a surplus of D + E + H, now earn a surplus of B + D. Thus, producers lose E + H − B. The net social loss to consumers and producers is C + G + E + H. The full set of market data is needed to estimate losses: the pre- and post-outbreak prices and quantities, as well as the slopes (elasticities) of the supply and demand functions.

The same resistance to changes in prices, or "price stickiness," that occurs when demand changes can also occur when supply changes (figures 4-3(d) and 4-3(e)). With producers unwilling or unable to raise prices, the quantity of goods and services supplied falls to Q_1 in figure 4-3(d). Thus, consumers lose a surplus of B + C. Producers originally earning a surplus of D + E + F earn only D after the outbreak-induced cost increase. Thus, producers lose a surplus of E + F. The net social losses are B + C + E + F. Note that this loss differs only slightly from the net loss shown in figure 4-3(c) where prices responded to changes in supply, the difference being the area B in figure 4-3(d).

In the final subcase, neither prices nor quantities change during the outbreak. This situation might occur if an outbreak began and ended so quickly that market forces did not have time to induce either a change in prices or a change in quantity; it is illustrated in figure 4-3(e) for an upward-sloping supply curve. As shown, prices and quantities remain fixed at P_0, Q_0. When supply changes from S to S', consumer surplus is unchanged at A + B. However, the higher costs of production now extend to all Q_0 units. The pre-outbreak producer surplus is D + E; the post-outbreak producer surplus is D − B − C; thus, producer losses amount to B + C + E. Note that no information on the demand curve is needed to calculate this loss. However, information on the pre- and post-outbreak supply functions is necessary.

Changes in Demand and Supply Because different combinations of assumptions concerning the shapes of demand and supply functions create an unwieldy number of cases when both shift simultaneously, only one of the more relevant cases, which appears to have wide application to the Luzerne County outbreak, is discussed. In this case there is a downward shift in a downward-sloping demand function and an upward shift in an upward-sloping supply function; prices remain unchanged. Figure 4-4(a) depicts the case where the outbreak causes a reduction in the quantities sold; by coincidence, prices remain the same. Welfare losses are represented by the area A + B + C + D + E.

The more realistic case of constant prices involves price stickiness, depicted in figure 4-4(b), where (in this example) demand decreases more than supply (that is, the price, P_0, intersects the demand curve *before* the supply curve). The new quantity demanded, Q_1, is found where the constant price intersects the new demand curve, D', assuming businesses meet demand but do not take advantage of the outbreak to maximize profits. The loss in consumer surplus is represented by areas A + B. Before the outbreak, producers earn a surplus of C + D + E + F + G + H. After the outbreak, producers earn only H. Thus, the loss in producer surplus is C + D + E + F + G. The total social loss is represented by all the shaded areas in figure 4-4(b).

A similar figure could be drawn for a case where demand shifts less than supply (where the price, P_0, intersects the supply curve *before* the demand curve). In both cases, the full range of market information, including pre- and post-outbreak prices and quantities and the slopes of the functions, is necessary to calculate the social losses.

From Theory to Application

No business sector is immune to all the losses associated with an outbreak of giardiasis, although some businesses may be affected less than others.

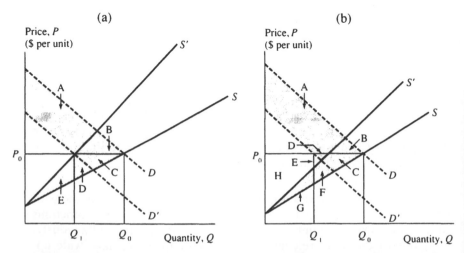

Figure 4-4. Losses in consumer and producer surplus due to simultaneous shifts in supply and demand: (a) price constant by coincidence; (b) price held constant.

Employees in any business may miss work due to illness or to caring for someone who is ill; ailing employees may be less productive than usual; and the steps people in the community take to avoid the disease may involve changes in the demand for goods and services. Changes in demand also involve losses.

Classical supply and demand theory, as developed in the previous section, provides a framework for the collection of data and the analysis of social losses resulting from an outbreak of giardiasis. This framework requires several steps to estimate losses. First, each type of business affected by an outbreak is categorized by one of the cases delineated in the previous section. Second, the available data are inventoried and compared with those specified by theory. Third, if a gap exists between the ideal data and reality, consideration is given to the collection of new data. Finally, the nature and direction of errors introduced by the use of imperfect or surrogate data are assessed.

Shifts in Supply Induced by Water Use   Businesses likely to be most affected by an outbreak of giardiasis use water to produce or provide food and drink. Restaurants and bars, which serve water and ice and use water to prepare food and wash dishes, are certain to be affected. Food markets, meat-packers, soda bottlers, and other commercial food handlers are also at risk, as are nursing homes, hospitals, schools, hotels, and motels. Those businesses that require pressurized water, such as dental services, face special problems because substitutes for tap water generally are not pressurized. Day-care centers incur additional risk from waterborne giardiasis because of the relatively high probability that the disease will spread once even a single child is infected.

As a first approximation, it is reasonable to assume that most businesses in an affected area do not raise prices. This assumption may seem strong at first, but firms selling in national, or even regional, markets may be unwilling or perhaps unable to adjust prices in response to changes in supply and demand. Even firms with some monopoly power, such as restaurants, bars, and hotels, may be reluctant to raise prices because they want to avoid both the loss of goodwill and what economists call transactions costs. To illustrate, a restaurant in an affected community might have some monopoly power because of its location or its unique bill of fare. From a narrow, profit-maximizing perspective, it could increase profits (or at least reduce losses) during an outbreak by raising prices. Increasing prices to reflect the higher costs of compliance with a boil-water order, however, would necessitate changing the menu, a step not only costly and temporary, but one that might erode goodwill and long-term profits.

Hospitals, nursing homes, day-care centers, and dentists may be more likely to raise prices than restaurants would. Hospitals and nursing homes may be in a monopoly position. Few small communities have more than one hospital or nursing home. And "brand" loyalty is likely to be strong in all four of these markets. Still, considerations of goodwill may be important in restraining price increases.

The situations facing pure competitors and monopolistic competitors are best represented by the supply and demand functions shown in figures 4-3(d) (for reductions in Q) and 4-3(e) (for constant Q), respectively. Cost increases induce the upward-sloping supply curve to shift upwards, but because the cost increases are not passed on as higher prices, the losses in social welfare either are shared by businesses and consumers if the quantities produced and sold change (figure 4-3(d)) or are absorbed by businesses alone if the quantities produced and sold do not change (figure 4-3(e)).

Shifts in Supply Induced by Labor   Lost work days and lost productivity of employees affected by giardiasis will raise production costs, and, at least in theory, will shift the supply curve upward, as shown in figures 4-3(d) and 4-3(e). The quantity of goods and services supplied may also fall. Some losses in production may be made up after the outbreak is over. Because only a small fraction of the residents in an affected community will be ill enough to miss work or to experience reduced productivity, supply curve shifts attributable to effects on labor are likely to be slight and possibly insignificant. Given this, and the price stickiness associated with considerations of goodwill, transactions costs, and outside competition, prices will probably not change. Thus, although lost work days and losses in productivity are counted as losses to individuals, we assume that supply price and quantity effects of such losses are zero.

Shifts in Demand   Reduction in local demand for goods and services is unlikely to result in price reductions, primarily because of the transactions costs. Franchise agreements may also limit the flexibility of local establishments to adjust to temporary changes in demand. Reductions in the quantity of goods and services demanded are likely to be small in any case, although there are some exceptions. Reductions in some leisure activities, such as going to the movies and eating out, may be large in isolated instances. Figures 4-2(b) and 4-2(c) both apply when the demand curve shifts but prices remain fixed: the former when the supply curve is assumed to be upward-sloping, the latter when it is assumed to be L-shaped.

When a giardiasis outbreak is announced, the most discernible increase in demand is likely to be for bottled water, anti-*Giardia* drugs, laboratory services, and related items. As stated earlier, however, effects that may be positive for bottled water dealers and pharmacists do not offset social losses.

Empirical Estimation

Despite the variety of cases that might apply to businesses affected by an outbreak, with few exceptions the theoretically ideal data for calculating social welfare losses in each case are identical: the slopes of the supply and demand functions (or the elasticities), and the quantities and prices of the goods traded, pre- and post-outbreak. Data on revenues may substitute for either price or quantity information.

Theoretically ideal data are not always available. The issue for analysis, then, becomes one of determining the consequences of using the data that are available. Prices and quantities before and during the outbreak can be obtained through interviews with businesses in the affected area. Demand elasticities are available from the economics literature for many products and services. In their absence, one can either use elasticities of a similar product, or bound the elasticities by referring to known elasticities of similar products for which consumers are likely to be less and more responsive to price, respectively.

Data on the supply side may be harder to obtain. If the supply elasticities before and during the outbreak are unavailable, data on the operating characteristics of average or typical firms, including profit rates, labor costs, depreciation, cost of goods sold, and other data appearing on accounting statements, may be used as substitutes. Such data may be available from publications that analyze the economic health of an industry.

The theoretical development in the first part of this chapter revealed that losses in producer surplus have two basic components: the costs of complying with the boil-water order and the revenue losses minus avoided costs associated with reductions in demand. The compliance cost is represented by the area E in figure 4-4(a) and the revenue losses minus avoided costs by the areas C + D. Data from industry studies are useful for estimating the latter cost.

Revenue losses minus avoided costs equal the change in profit before taxes. Thus, only data on gross profits are needed to value the second component. However, reported profits before taxes are usually net of accounting depreciation, which usually bears little relationship to "real" depreciation—the actual wear and tear on business assets. Thus, the "accounting" portion of depreciation needs to be added back to reported profits before the change in "real" profits is computed.[5]

The data presented in industry studies, such as reported profits and accounting depreciation, are likely to be applicable to an average firm making decisions over time, not to a group of firms in the unusual and short-run business circumstances associated with an outbreak of giardiasis. The average firm in normal circumstances is likely, for example, to have more flexibility in labor and pricing decisions than those firms in an outbreak area. If that is so, the calculations of short-run losses based on profit rates associated with an average degree of employer flexibility and discretion in avoiding costs will underestimate losses in producer surplus in an outbreak area. Had the firms affected by the outbreak had more discretion and flexibility, they would have taken steps to cut their losses further. (This point assumes full employment in the economy. See below for a qualification.)

Further Considerations

The theoretical discussion of welfare losses associated with an outbreak of giardiasis has avoided a number of important issues relevant to benefit estimation. The first concerns the size of the geographical area defined as *society* in the phrase *social cost*. In general, the larger the geographical area considered by the study, the more a given welfare loss in the community with contaminated water will be offset by gains elsewhere in the area. If tourists avoid a community experiencing an outbreak, the local economy will certainly suffer, but other communities may gain much of this business. In that event, the losses to society will be less than those indicated by a more geographically restrictive study.

Just as losses in one area may be offset by gains in another, losses in one time period may be offset by gains in future periods. If, for example, ill people avoid dining out during their bout with giardiasis but make up the lost opportunities when they are well, business losses (as well as losses to society) taken over a longer period will be lower than if those dining opportunities were permanently "lost."[6]

In tallying social losses to businesses in an affected area, two other issues need to be addressed. The first concerns the effect of idle resources; the second, the treatment of investment goods.

---

[5]If all depreciation is added back to profits, however, losses will be overestimated, because some wear and tear will be avoided when output is reduced. This analysis assumes that real depreciation is zero during the outbreak.

[6]We have argued above that post-outbreak demand for food and drink is unlikely to rise by much.

Idle Resources Assume that a restaurant lays off workers because an outbreak decreases demand for its service. The restaurant saves the costs of wages (and other variable costs), which would be subtracted from the loss in revenue to calculate loss in producer surplus. If the unemployed workers do not find immediate employment, the productivity of their labor is "lost" to society. (Productivity gains in a future period may reduce this loss, perhaps at some increase in cost.) In this case, the labor costs avoided are real costs to society and should be included in the social losses. However, if the layoff occurs under conditions of full employment, the workers would quickly find new employment and their productivity would not be "lost" to society. In this case, the loss in social welfare is smaller, consisting of lost revenues minus the nonlabor costs avoided.

One further complication is involved with layoffs. It is reasonable to assume that workers seek to maximize total income, pecuniary plus nonpecuniary returns. Even if alternative employment is obtained, it is unlikely to provide the same total income as the worker's former job. This difference in total income should be considered a social loss even under the full employment assumption.

The RFF Restaurant and Bar Survey, reported in chapter 7, revealed that outbreak-related layoffs were rare. Because of that finding and because the complications of considering layoffs far outweighed the benefits in improved estimates, they were ignored in the loss estimates.

Investment Goods Investment by businesses in goods to avoid contamination presents special problems in computing producer surplus losses if the investment goods provide benefits to society after the threat of contamination ends. In this case, the social costs and benefits are spread over many years. The treatment of this cost requires an explanation.

Let M denote the net present value of the costs of water supplied over the lifetime of the investment by municipal or other sources not under the firm's control. Let N be the net present value of the costs of water obtained from sources under the firm's control, such as a well or a treatment plant that removes the contaminant. Assume that before the outbreak, $M < N$ for all profit-maximizing firms with "external" sources of water supply.[7] A firm that invests in an alternative source of

---

[7]Russell, Arey, and Kates (1970) found that the present discounted cost of alternatives to municipal water was actually less than the present discounted cost of municipal water for many firms that were connected to the municipal water supply system. They found that firms were not profit-maximizing and that there was a large degree of "slack" in their adoption of water-use technology. If that is true for firms in the case study area, the adoption of such technology may actually be a social benefit and not a cost at all.

water to avoid contamination incurs a net present value cost of $N - M$. Depending on the likelihood and frequency of future contamination, only a portion of this cost should be allocated to the first outbreak. Because the information needed to make this allocation was unavailable, investment costs that provide long-term benefits were ignored, which biases our loss estimates downward. However, interviews revealed that few individuals or businesses made investments to protect themselves from the outbreak, so this omission should have little effect on the estimates.

LOSSES TO GOVERNMENT AGENCIES

Government and community involvement in an outbreak of waterborne giardiasis is likely to be triggered when an unusually large number of cases of gastroenteritis are reported to public health authorities. Authorities must quickly find answers to several questions: What is the pathogenic organism causing the illness? Through what medium—food, water—are people being exposed to it? How is the organism getting into this medium? How extensive is the outbreak?

To answer these questions, public health authorities at the state level initiate a number of investigations, sometimes with local support or help from the Centers for Disease Control in Atlanta or the EPA. First, they conduct, usually by questionnaire, an epidemiological survey of the community and of one or more control communities nearby. This study reveals the extent of the outbreak and the range of symptoms experienced. It also may give clues to the source of the pathogen. Often a subset of this group is requested to submit stool samples to be examined for various organisms. These laboratory tests are designed to confirm the cause of the outbreak, although for nearly half of all outbreaks of gastroenteritis, no cause is ever established (Centers for Disease Control, 1982).

Because outbreaks of gastroenteritis are often waterborne, a special survey may also be undertaken to compare sicknesses in households served by the suspect water supply system with sicknesses in nearby households served by other water supply systems. If the water supply is implicated, and if the pathogen has been identified in the epidemiological survey, the water supply itself—both the raw and finished water—is examined for pathogens. The water treatment system may also be examined, assuming there is one, to determine how viable pathogens entered the finished water. The water treatment and water quality examinations are usually performed in cooperation with the water supply utility.

From our interviews with government officials at several outbreak sites, it appears that government authorities and the water utility may spend considerable time negotiating an appropriate response to the outbreak, particularly if a private water utility is involved. This time represents a cost to society because it is time that could, in principle, be spent on other productive activities.

Government authorities at all levels may collaborate to develop a strategy to help citizens avoid the disease. During the outbreak in Luzerne County, for example, the Emergency Management Agency in Wilkes-Barre was brought in to coordinate a response to the crisis that involved two state agencies, the Army National Guard, a number of community groups, and several municipal governments. Even an operation seemingly as simple as distributing clean, free water requires that tank trucks or other vessels be obtained, storage space for the tank trucks be found, and personnel to oversee the distribution of water be lined up. Particular problems may be encountered in meeting the water needs of the handicapped and the elderly.

Separate steps are usually taken to alert restaurants, meat-packers, and other food handlers as well as other businesses that expose the public to water, such as dentistries and suppliers of ice. Meetings with the press and concerned citizens are often an essential component of informing the public about an outbreak. Inspections usually accompany boil-water orders.

If businesses decide to recover damages from the outbreak through the court system, the legal fees and court fees incurred are also social costs; the settlement, though, is not a social cost but a monetary transfer from the defendant to the plaintiffs. If the water utility is privately owned, the state public utility commission may meet to decide on the costs the utility will be allowed to recover.

The valuation of government costs is relatively straightforward if prices (including wages) are assumed to reflect marginal social costs. In that case, the salaries earned by public officials and employees when working on an outbreak and the associated costs of travel, food, and lodging are the social losses due to the outbreak. Thus, in valuing time at the wage rate, it is assumed implicitly that government employees and lawyers involved in litigation and other aspects of the outbreak push aside equally valuable tasks to contend with the outbreak.

This assumption raises a question: Does the time government employees spend on an outbreak have an opportunity cost? There is no question that it does for overtime work (see chapter 3). The controversy concerns normal work time. From a long-run perspective, all time has an opportunity cost. If emergency events, such as fires and outbreaks of giardiasis, were to become less likely or less frequent, the number of employees in these agencies could be reduced. Those employees laid off would take productive jobs elsewhere—in effect, the jobs they "pushed aside" when they chose government employment. From a short-run perspective, however, counting normal work time spent on an outbreak as a social loss is more debatable because salaries would be paid with or without the outbreak. Morever, the employee might not have other tasks to "push aside." In estimating the losses to government agencies, we take the long-run perspective because our ultimate concern is with the benefits that accrue from a long-run investment—the construction of a water treatment plant. This issue is discussed in more detail in chapter 8.

Another complicating factor is how to value the time that unpaid volunteers spend on an outbreak, distributing water to the elderly, for example. One way is to use the alternative cost of supplying the same service through the market. This procedure is analogous to that used to value unpaid household services.

A final complicating factor, which is ignored in the case study, is the institutional learning—the investment in human capital—that takes place during a crisis. To the extent that it will help in a future crisis, this learning is a benefit of the outbreak, not a cost. Analogous to the procedure used to handle real investment in response to the outbreak, the time costs spent in coordinating emergency responses among the different government agencies should not be allocated solely to the outbreak under study if such investments will reap dividends during future crises. Because the elements of public costs that will be of benefit in future crises are difficult to identify, all government time costs are allocated to the outbreak under study, a procedure that biases these costs upward.

LOSSES TO THE WATER SUPPLY UTILITY

The water supply utility is likely to face considerable costs if its finished water is contaminated. The time the utility spends formulating a response to the crisis and meeting with community and state officials, citizen groups, the public utility commission, and the press is time not spent on normal duties and responsibilities. In addition, the company is likely to take emergency steps to reduce the risk of further contamination and illness, such as constructing pipelines to divert water from uncontaminated sources, establishing and maintaining free water distribution centers, using higher levels of chlorine than normal in the water supply, and increasing the frequency of water quality testing.

Valuing the social costs of work time is relatively straightforward if wages are assumed to reflect marginal social costs. Expenditures on wages for time allocated to public relations and to the formulation and implementation of the water utility's emergency responses to the contamination are both considered to be social costs.

Valuing the social costs of emergency steps taken to reduce the risk of further contamination is also straightforward, assuming that prices reflect marginal social costs. Here, as for all capital investments, the benefits of measures taken to protect the public may extend beyond the period of contamination, and therefore, total investment costs should not be attributed to a particular outbreak. For example, a pump purchased to divert water during the contamination period becomes an integral part of the water distribution system even after the outbreak has ended. Therefore, only a portion of the investment should be counted as a social cost of the particular outbreak.

REFERENCES

Centers for Disease Control. 1982. *Water Related Disease Outbreaks Annual Summary 1981*, U.S. Department of Health and Human Services, Health and Human Services Publication (CDC) 82-8385 (Atlanta, Ga., U.S. DHHS).

Russell, Clifford S., David G. Arey, and Robert W. Kates. 1970. *Drought and Water Supply: Implications of the Massachusetts Experience for Municipal Planning* (Baltimore, Md., Johns Hopkins University Press for Resources for the Future).

Willig, Robert D. 1976. "Consumers' Surplus Without Apology," *American Economic Review* vol. 66, no. 4, pp. 589–597.

Part 2 / Luzerne County Case Study

5 / The Luzerne County Outbreak

Before the methods developed in chapters 3 and 4 are applied to estimate the losses associated with a specific outbreak of giardiasis, it is useful to describe that outbreak and the responses to it. This chapter describes the area that was affected by the outbreak, lays out the events surrounding the outbreak, and discusses actions taken by individuals, businesses, government agencies, and the water supply utility to contain the outbreak, avoid the disease, and restore a safe drinking water supply for the affected area. A chronology of major developments during the course of the outbreak is presented in appendix 5-A.

THE AREA AFFECTED BY THE OUTBREAK

An increase in the incidence of giardiasis among residents of several small communities near Wilkes-Barre, Pennsylvania, during the late fall of 1983 was linked to drinking water supplied from the Spring Brook Intake Reservoir, which is located in Lackawanna County and owned and operated by the Pennsylvania Gas and Water Company (PG&W). In an effort to contain the spread of the disease, PG&W and the Pennsylvania Department of Environmental Resources on December 23 advised 75,000 consumers living in eighteen communities to boil their drinking water. By mid-1984, 370 confirmed cases of giardiasis had been reported to the Pennsylvania Department of Health, which, at the time, made the outbreak one of the largest ever recorded in the United States in terms of confirmed cases.

The affected area stretches along the east and west banks of the Susquehanna and Lackawanna rivers between Wilkes-Barre and Scranton in upper Luzerne and lower Lackawanna counties (figure 5-1). Ringed by rolling Appalachian Mountains, the eighteen affected communities lie on the floor of the Wyoming Valley within the greater Wilkes-Barre metropolitan region. The financial and commercial center of Luzerne County, Wilkes-Barre (population 59,000) also boasts manufacturers of iron, textiles, apparel, and perfume.

The total population of the eighteen affected communities was 100,290 in 1980 (U.S. Bureau of the Census, 1982). However, only 75,000 residents were placed on the boil-water advisory. These residents were located in the Spring Brook–Hillside Service Area, an integrated water distribution system with two sources of supply, one that turned out to be contaminated and one uncontaminated. These two sources—the contaminated Spring Brook Intake Reservoir (with no filtration prior to distribution) and Huntsville Reservoir (with a filtration plant at Hillside)—are shown in figure 5-1. Approximately 25,000 other residents in the affected communities were supplied with water from other uncontaminated sources. A list of the communities affected by the outbreak, the number of people in each community, the number of people in the Spring Brook–Hillside Service Area affected by the boil-water advisory, and the distribution of the affected population among four age groups are provided in table 5-1.

To fully understand the outbreak and the actions PG&W took to deal with it, some background on the region's water supply and distribution system is helpful. At the time of the outbreak, PG&W supplied water to approximately 600,000 people living in parts of Wayne, Susquehanna, Luzerne, and Lackawanna counties. The extensive water supply system included fifty-five storage reservoirs, of which twenty-seven were in service. The company's water service territory was divided into two rate areas, Spring Brook, which served principally Luzerne County and a small portion of Lackawanna County, and Scranton, which served most of Lackawanna County and small portions of Susquehanna and Wayne counties. All eighteen communities affected by the boil-water advisory fell within the Spring Brook Rate Area. Water in the affected area was supplied through two intake reservoirs, Spring Brook Intake Reservoir (elevation 910 feet) and Huntsville Reservoir (elevation 1,134 feet).

Spring Brook Intake Reservoir, located in Spring Brook Township in Lackawanna County, had a storage capacity of 25 million gallons and a drainage area of 6.14 square miles. It was supplied primarily by Nesbitt Reservoir (elevation 1,155 feet), which has a capacity of 1,279 million

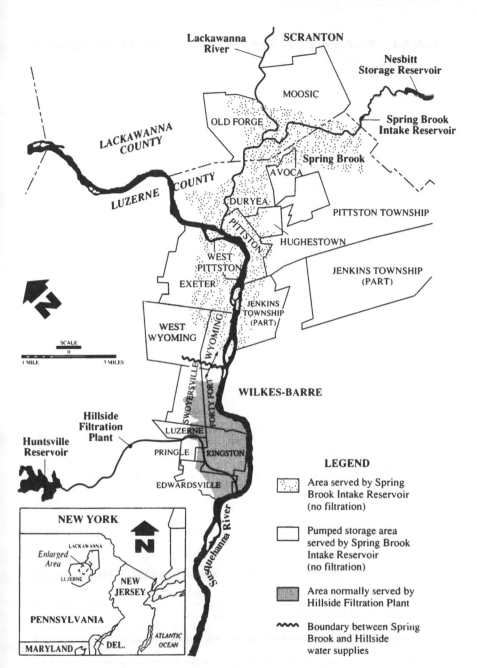

Figure 5-1. The eighteen communities, located between Wilkes-Barre and Scranton, affected by the boil-water advisory of December 23, 1983.
Source: Based on information provided by the Pennsylvania Gas and Water Company, Wilkes-Barre, Pennsylvania.

Table 5-1. Population in Case Study Area Affected by Boil-Water Advisory of December 23, 1983

Municipality	Total 1980 population[a]	Population estimated to be affected by advisory			
		Total population[b]	Preschool (under 5 years)[c]	School age (5–18)[c]	Working age (19–64)[c]	Retired (65 and over)[c]
Avoca Borough	3,536	1,132	60	242	658	172
Duryea Borough	5,415	5,337	279	1,023	3,111	924
Edwardsville Borough	5,729	1,710	119	340	967	284
Exeter Borough	5,493	5,433	309	1,194	3,140	790
Forty Fort Borough	5,590	5,590	299	1,169	3,160	962
Hughestown Borough	1,783	879	46	213	508	112
Jenkins Township	4,508	66	4	14	37	11
Kingston Borough	15,681	15,681	851	2,772	8,771	3,287
Luzerne Borough	3,703	1,921	94	338	1,099	390
Moosic Borough	6,068	2,307	147	514	1,391	255
Old Forge Borough	9,304	7,722	398	1,430	4,547	1,347
Pittston City	9,930	9,930	563	1,968	5,705	1,694
Pittston Township	3,611	20	1	6	11	2
Pringle Borough	1,221	196	8	48	111	29
Swoyersville Borough	5,795	4,003	172	819	2,366	646
West Pittston Borough	5,980	5,980	314	1,183	3,419	1,064
West Wyoming Borough	3,288	3,288	153	718	1,979	438
Wyoming Borough	3,655	3,655	147	602	2,116	790
Total	100,290	74,850	3,964	14,593	43,096	13,197
Percent		100	5.3	19.4	57.7	17.6

Note: Table is based on the population residing in the affected area in 1980.

[a]From Bureau of the Census, *1980 Census of Population* vol. 1, chap. B, *General Population Characteristics, Part 40, Pennsylvania.* Report PC 80-1-B40, August (Washington, D.C., Government Printing Office, 1982).

[b]Estimated by Pennsylvania Gas and Water Company, Wilkes-Barre, Pennsylvania. Personal communication, Gerald B. Taylor, Vice President, Operations and Engineering, June 19, 1984.

[c]Based on proportion of total population in these age groups for individual cities, boroughs, towns, and townships in the affected area.

gallons and a drainage area of 20 square miles. At the time of the outbreak, water from the Spring Brook Intake Reservoir was chlorinated before distribution but not filtered.

Huntsville Reservoir, in Lehman Township in Luzerne County, had a storage capacity of 1,922 million gallons and a drainage area of approximately 8.4 square miles. Water from Huntsville Reservoir was filtered and chlorinated at the Hillside Filtration Plant before distribution.

Water from both reservoirs was delivered to the Spring Brook–Hillside Service Area through a gravity feed distribution system. (West Wyoming Borough, one of the affected communities, was served by a pumped storage system supplied primarily by water from Spring Brook Intake Reservoir.) Because of the integrated distribution system, the entire service area technically could be supplied by either reservoir, but under normal circumstances, most of the affected communities were served by one or the other reservoir. A zone dividing the two parts of the service area was served by both reservoirs, depending on the directions of the pressure gradients in the water mains. The areas served by the two reservoirs on December 23, 1983, are shown in figure 5-1, as is the normal location of the boundary between the Spring Brook and Hillside sources of water supply.

After the boil-water advisory was issued and the Spring Brook Intake Reservoir was implicated as the source of *Giardia* cysts, PG&W devised a plan to divert water to some of the affected area from uncontaminated sources of supply outside the Spring Brook–Hillside Service Area. By closing existing valves, installing new ones, and disconnecting and capping several water mains, PG&W physically separated those sections of the distribution system that could be supplied with water from uncontaminated sources from those that could not be. The latter had to remain on the contaminated Spring Brook supply.

The new divides in the water distribution system permitted the boil-water advisory to be lifted by June 1, 1984, for more than 70 percent of the customer accounts originally affected, leaving only 7,803 accounts, affecting approximately 21,000 residents, under the advisory. (The number of customer accounts affected by the advisory, and the dates they were released from it, are shown in table 5-2.)

Those accounts remaining under the advisory after June 1 could not be provided with an alternative source of water because the normal demands for water exceeded the capacity that could be handled by the existing water mains without adversely affecting flow and pressure. (Fire protection requires adequate pressures and flows, and public health considerations require positive pressures throughout the distribution system.) The 7,803 accounts continued to receive water from Spring Brook Intake Reservoir until September 21, 1984, when PG&W began

Table 5-2. Number of Accounts in Spring Brook–Hillside Service Area Affected by Boil-Water Advisory, and Duration of Advisory

Type of account	Total number of accounts on advisory[a]	Date in 1984 when boil-water advisory was lifted					
		March 30 (On advisory 99 days)	April 9 (On advisory 109 days)	April 20 (On advisory 120 days)	May 9 (On advisory 139 days)	May 26 (On advisory 156 days)	Sept. 21 (On advisory 274 days)
Residential	25,701	13,166	2,163	31	1,627	1,464	7,250
Commercial	1,861	1,023	129	28	94	92	495
Industrial	138	72	9	0	5	5	47
Municipal	64	44	2	0	4	3	11
Total	27,764	14,305	2,303	59	1,730	1,564	7,803
Cumulative percent of total accounts		51.5	59.8	60.0	66.3	71.9	100.0

Source: Pennsylvania Gas and Water Company, Wilkes-Barre, Pennsylvania. Personal communication from Gerald B. Taylor, Vice-President, Operations and Engineering, June 19, 1984.
[a]Most accounts were placed on a boil-water advisory on December 23, 1983. Accounts in Pringle Borough and sections of Edwardsville Borough were not added to the advisory until January 6, 1984.

using a new 15,000-foot pipeline from Nesbitt Reservoir located 1.5 miles upstream from Spring Brook Intake Reservoir (figure 5-1).

THE FOUR STAGES OF THE OUTBREAK

As outlined in chapter 2, there are four stages in investigating any outbreak of a waterborne disease: discovery, study and testing, reaction, and aftermath. The following sections describe those four stages in the Luzerne County outbreak.

The Discovery Stage

The discovery stage occurs when health authorities at the local or state level receive information on the incidence of confirmed cases of giardiasis suggesting that an outbreak of the disease may be in progress. During the first two weeks of October 1983, officials at the Pennsylvania Department of Health noted that area physicians were reporting an unusually large number of cases of giardiasis—well above the normal endemic rate of one case per month per township or borough. Recognizing the possibility of an epidemic, the Department of Health wrote to physicians throughout the greater Wilkes-Barre area in early November, informing them of the recent increase in the incidence of giardiasis and advising them to report any new cases they might discover. From late November through December, the number of cases of giardiasis reported to the Department of Health's district epidemiologist in Kingston shot up, leading state and local health departments to begin an in-depth investigation into the cause of the disease.

The Study and Testing Stage

Once a significant number of confirmed cases of giardiasis have been discovered in an area, state or local epidemiologists, the Centers for Disease Control in Atlanta, or EPA personnel begin conducting epidemiological surveys to confirm the presence of an outbreak, ascertain its size, and isolate its cause. Such a survey marks the beginning of the study and testing phase.

In an effort to pinpoint the source of the giardiasis outbreak in Luzerne County, the district epidemiologist in Kingston in December 1983 sent a questionnaire to each person with a confirmed case of the disease as the case was reported. The questionnaires asked patients for a history of their water consumption, including the sources of their drinking water;

for a history of their illness, including symptoms and treatment; and for information concerning other household members who may also have experienced a diarrheal illness within the previous three months. The only common element among the first giardiasis victims to complete their questionnaires seemed to be consumption of bottled water. As the investigation continued, however, attention began to focus on the municipal water supply as the probable source of the disease. By late December 1983, the one common denominator among the vast majority of confirmed cases of giardiasis was indeed the Spring Brook Intake Reservoir. The State Department of Environmental Resources notified PG&W of this development at a meeting on December 22, 1983. The next day, PG&W and the state agency jointly issued a boil-water advisory for all consumers of water from Spring Brook Intake Reservoir.

The next priority for the Department of Health was to ascertain the size of the outbreak through a random telephone survey of the residents in the area. On December 27 and 28, 410 households, with 617 residents, were queried on the size of the household and recent diarrheal illnesses within the household, including date of onset, diagnosis, and treatment, if applicable. The sample included consumers of water from Spring Brook Intake Reservoir and from other PG&W reservoirs in the area. Defining a clinical case of giardiasis as a diarrheal illness with acute symptoms lasting ten days or more, state epidemiologists recorded twenty-one incidents of giardiasis-like illness among a sample of 233 residents served by water from Spring Brook Intake Reservoir, a gross attack rate of 9 percent (21/233 = 0.094). Compared with the 1 percent attack rate found in individuals not served by water from Spring Brook Intake Reservoir, the finding further implicated this reservoir as the source of giardiasis.

While the epidemiologists were surveying area residents, the EPA, called in by the state environmental agency, was testing the suspect reservoir for *Giardia* cysts. Five days after the boil-water advisory was issued, the EPA took samples of both raw and finished (chlorinated) water from Spring Brook Intake Reservoir and samples of sediment from the reservoir bed. Two weeks later, on January 12, 1984, the test results were publicly announced: *Giardia* cysts were recovered in the majority of samples taken from both raw and treated water from Spring Brook Intake Reservoir, eliminating any doubt that the reservoir was the source of the outbreak.

A physical survey of watershed lands was also made. PG&W and local residents initially believed that beavers inhabiting the area were the actual source of the disease and had contaminated the reservoir. Nine beavers trapped near the Spring Brook Intake Reservoir were autopsied, and two were found to be *Giardia*-positive. The Department of Environmental Resources, however, believed that human sewage from poorly constructed residential septic systems and a malfunctioning wastewater treatment plant located in Roaring Brook Township were the ultimate sources of *Giardia* cysts. The agency believed that the beavers merely served as carriers of the disease, picking up the cysts from surrounding tributaries and streams contaminated with human sewage and transporting them to the intake reservoir.

The Reaction Stage

During this stage, local or state authorities may issue advisories or orders to boil water to minimize exposure to the contaminated water supply; administrative, legislative, and judicial inquiries may be initiated into the nature and cause of the outbreak; and temporary measures may be taken to repair the water supply system, treat the water supply, and reduce the risk of recontamination to acceptable levels. Nearly all of these steps occurred during the reaction phase of the Luzerne County outbreak.

Reaction of Government Agencies As already reported, the Department of Environmental Resources and PG&W issued a joint boil-water advisory to all customers residing in the Spring Brook–Hillside Service Area on December 23, 1983. Residents were told to heat their tap water to a rolling boil for one to three minutes before using it for drinking or in food preparation.

At the same time, the state agency ordered all public eating and drinking establishments within the affected area to use water approved for drinking (either boiled or bottled) in all water-related food preparation, including salad bar preparation and beverage mixing (tea, coffee, lemonade, and soda). Ice used in drinks or in contact with food also had to be made from water approved for drinking, since *Giardia* cysts can survive freezing temperatures. A copy of the announcement of this order is included as appendix 5-B to this chapter.

Local and county governments reacted immediately to the boil-water advisory. Local governing bodies had to decide whether to provide potable water for their residents, and to take on all the time, expense, and administration such a decision would entail, or to encourage residents to comply with the boil-water advisory. In the short run, most communities were forced to choose the second alternative, although the few who received canned water donated by an area brewery in the first few weeks of the outbreak were able to avoid the decision temporarily. In the end, most municipalities devised plans to request portable, 450-gallon tanks, known as water buffaloes, from the Army National Guard and to operate them in cooperation with PG&W.

To avoid freezing, the tanks had to be kept indoors, usually in municipal buildings, garages, or local firehouses. Some municipalities had nowhere to house the tanks and so could not use them. Those municipalities that did use them had to supervise the filling of buffaloes and the distribution of water to local residents. After the first month, some of the communities' manpower needs were eased when PG&W purchased a tanker to transport water directly to distribution centers. Nonetheless, some communities decided not to establish distribution centers because they could not afford even the minimal expense for the necessary manpower. Other communities declined help because they did not trust the quality of the local water. (This concern was justified to some extent; two distribution points were supplied with water from Elmhurst Reservoir near Scranton, which was later found to be contaminated with *Giardia* cysts.)

The Luzerne County Emergency Management Agency kept detailed records of the quantity of water distributed each week from the various buffaloes. These data show that while most communities implemented a temporary water distribution system, many residents in the area failed to take advantage of it. Kingston's nearly 16,000 residents, for example, used an average of 600 gallons of tanker water per week, a mere 4.8 ounces per resident per week. Residents of West Pittston, on the other hand, consumed more than 5,000 gallons of tanker water per week, approximately one-half gallon per person per week.[1]

The drinking water needs of the area's elderly and handicapped population were addressed by both local and county governments. At the county level, a task force, coordinated by the Emergency Management Agency and made up of the Office of the Aging, the Office of Human Services, the Commission on Economic Opportunity, and the Transportation Authority, used personnel from the county's Housing Authority to obtain water from PG&W distribution centers and deliver it in small, manageable containers to the three apartment complexes for the elderly in the area. The Emergency Management Agency bought 200 of these containers for three cents each from Army surplus.

In addition, several boroughs also arranged to deliver water to their elderly residents and shut-ins. In West Pittston, for example, borough workers using borough equipment delivered water to about seventy shut-ins, who received at least 1 gallon every week.

Various units of state government conducted investigative hearings on the outbreak. In late March 1984, the Pennsylvania State House Conservation Committee held several hearings in the Wilkes-Barre area and gathered testimony from municipal officials, the Pennsylvania Public

---

[1]Personal communication from James W. Siracuse, Luzerne County Emergency Management Agency, Wilkes-Barre, Pa., June 19, 1984.

Utility Commission, and PG&W. A month later, a Pennsylvania Senate Committee on Water Quality heard testimony from the general public, local government officials, and PG&W. The Senate hearings focused on the effectiveness of the Department of Environmental Resources in policing drinking water quality and on the quality of finished water from treatment plants operated by PG&W.

The Public Utility Commission also initiated an investigation into the events surrounding the outbreak. From nineteen days of evidentiary proceedings and three days of public hearings, the commission hoped to (1) establish the water company's liability in the outbreak, (2) determine some of the direct costs associated with the outbreak, (3) determine the possibility of an immediate rate reduction or credit for affected PG&W customers, and (4) determine who would pay for PG&W's short- and long-term solutions to the contaminated-water problem. On July 20, 1984, the commission ordered the water company to reduce water bills of affected customers by 25 percent, retroactive to the beginning of the outbreak. The average refund check for residential users totaled $10.20 for each quarter the public water supply was considered unsafe to drink.

Reaction of the Water Utility   After the boil-water advisory was issued, PG&W began to search for alternative sources of safe drinking water and for means to ensure a safe water supply from Spring Brook Intake Reservoir.

In an effort to kill the *Giardia* cysts, PG&W doubled the chlorine level at the Spring Brook Intake to 4 milligrams per liter. At the same time, the company installed a continuous monitoring device to measure residual chlorine levels. For at least two weeks in early February 1984, customers complained of excessive levels of chlorine in their tap water. Although few of the reports were substantiated, residents told RFF interviewers stories of babies breaking out in rashes after their baths, of greenish chlorine stains on porcelain tubs and sinks, of family members unable to take showers because of chlorine fumes, and of numerous laundry loads bleached beyond repair by the excessive chlorine. Similar stories were reported in the media. (See, for example, *Wilkes-Barre Times Leader*, 1984).

Pennsylvania Gas and Water also undertook other short-term measures that proved less controversial. To remove one link of the chain of giardiasis transmission, the company, in cooperation with the Pennsylvania Game Commission and the Department of Environmental Resources, trapped and relocated several colonies of beaver residing in the Spring Brook watershed. PG&W also retained an independent engineering firm to examine the Spring Brook watershed for existing or potential sources of contamination and hired two expert consultants on the transmission of *Giardia*.

PG&W and local and county governments recognized that not all customers in the affected area had the facilities or the desire to boil their drinking water and that some did not trust the safety of boiled water. PG&W therefore provided an alternative source of water for its customers, setting up both small-container and 500-gallon tanker refill points at its two area service centers and at several of its local offices. The company encouraged municipal governments to use the centers, free of charge, to fill their tankers for distribution to local residents. In time PG&W purchased a large tanker to transport water to those municipalities that had water buffaloes. Deliveries were made once a day to individual distribution points located throughout the outbreak area.

At the same time, PG&W engineers were busy improvising a way to divert water from uncontaminated sources of supply within the PG&W system to areas ordinarily served by the contaminated reservoir. This scheme required physically separating those sections of the distribution system that could be supplied with water from uncontaminated sources from those that could not be and, thus, had to remain on the contaminated supply. PG&W spent more than $37,000 for the installation of new valves to create the necessary divides.

Once the divides were in place, sixty days had to pass while *Giardia* cysts were flushed from the distribution system. Only then could the boil-water advisory be lifted. Advisories were lifted in March, April, and May 1984 for more than 70 percent of the customer accounts originally affected by the advisory (see table 5-2). By June 1, only 7,803 accounts affecting approximately 21,000 residents remained on the advisory, which was lifted on September 21, 1984, when PG&W began using a new, $1.5-million, 15,000-foot pipeline from the Nesbitt Reservoir located 1.5 miles upstream from the Spring Brook Intake Reservoir (see figure 5-1).

Reaction of the Public Coming as it did just two days before Christmas, the boil-water advisory dampened the normal holiday cheer, as thousands of Wyoming Valley families were suddenly faced with the immediate prospect of boiling or buying water to prepare and serve their usual holiday feasts. Prices doubled for bottled spring water, which disappeared from store shelves as fast as it could be restocked.

As anger and frustration subsided and the holidays passed, people began to adjust their behavior to avoid the disease. Many residents boiled tap water for at least one minute to kill the cysts. Once cooled, the water could be transferred into containers of manageable size — indeed, plastic 1-gallon milk containers soon became a sought-after commodity—and used for drinking, food preparation, dishwashing, and personal hygiene. Some residents told interviewers they found brushing their teeth and washing salad greens particularly irksome activities during the outbreak.

Many residents also bought bottled water, either to replace or to supplement boiled tap water. Despite the increased demand for bottled springwater, few shortages were reported after the first week or two, and prices soon stabilized. Those who bought bottled water reserved it primarily for drinking.

A third option was to haul uncontaminated water from one of a variety of sources. Friends, relatives, employees, and employers in areas not affected by the advisory proved to be valuable sources of safe water for many under the advisory. Wells were sources for some, as were the dozen or so water distribution centers scattered throughout the outbreak area. There were, however, reports of long lines at the distribution centers in the first few days. Within a week or two, though, the lines shortened and the average wait dropped to between five and fifteen minutes.

Residents could also purchase and install home filtration systems, thereby avoiding boiling tap water or buying bottled water. Within a week of the outbreak, several advertisements appeared in local newspapers advocating the advantages and efficiency of home filtration systems. However, neither the Department of Environmental Resources nor the Department of Health would guarantee the effectiveness of any of the devices being advertised, and both agencies warned that they knew of no home filtration system that was totally reliable in filtering *Giardia* cysts. Given the associated risk, the expense, and the lack of public health endorsement, few people in the outbreak area chose to buy these systems.

Reaction of Businesses and Public Services Businesses and public services were not equally inconvenienced by the boil-water order. Most businesses merely had to disconnect their water fountains, but businesses with critical water usage faced a long and often costly ordeal.

Restaurants and bars were the hardest hit. The Department of Environmental Resources estimated that there were at least 212 licensed restaurants and taverns within the boil-water order area, all of which had to find an alternative supply of safe water for use in preparing beverages and food. The vast majority of restaurants stopped serving drinking water to customers unless requested. One restaurant and bar in Edwardsville Borough hauled uncontaminated water from a PG&W service center some fifteen minutes away. Twice a day busboys went to the distribution center to fill six 5-gallon containers and ten soda fountain canisters with water. A local Wendy's Restaurant, on the other hand, chose to boil water for use in food preparation, beverage making, and dishwashing. (There was no dishwasher on the premises.) The boiled water was transferred to two 50-gallon plastic drums purchased especially to store the water. Water still hot from boiling was used to wash dishes. The water stored in drums was used to make coffee and tea and to wash fresh vegetables for salads.

Franklin's Family Restaurant in Edwardsville exercised yet another option, obtaining safe water from any one of three sister franchise operations located outside the affected area, but within twenty minutes of the Edwardsville restaurant. Company trucks and restaurant employees were sent for the water, a labor-intensive activity that took time away from the employees' normal duties. Franklin's used the transported water to make coffee, tea, mixed beverages, and nondairy creamer; to wash fruits and vegetables; and to rinse food and drink containers that could not be put in a dishwasher.

Finally, several taverns, which do not normally serve or use as much water as the average restaurant, purchased bottled water for their customers. The taverns interviewed by RFF used an average of 40 gallons of bottled water a week.

Finding an alternative to tap water used in soda fountain systems proved to be a major inconvenience and cost for most of the restaurants and taverns affected by the boil-water order. Before the outbreak, Wendy's, for example, normally bought soda in concentrated form, to be mixed with carbon dioxide and water before serving. During the boil-water order, the restaurant switched to a form of soda that already included water and needed only to be mixed with carbon dioxide. Because 1 gallon of the concentrate equals 10 gallons of the soda mixed with water, Wendy's had to purchase about ten times more soda (by volume), at a higher net cost, than it would have if the water supply had been safe.

Other restaurants and taverns substituted bottled or canned soda for fountain soda, which more than doubled the cost per glass for several of the operations. Taverns and bars were especially hard hit because six of the twelve basic bar mixes used soda water.

The boil-water order also required that ice used in drinks and in contact with food be made from water approved for drinking. Franchised restaurants were able to obtain ice from sister operations outside the boil-water area, but even then, some had to buy additional ice to accommodate heavy weekend demand. Those restaurants that had no choice but to purchase ice made an effort to conserve its use in beverages. Conservation, however, meant using more liquid to fill a glass—a trade-off with its own costs. Bar owners experienced a similar problem. Ice cubes manufactured for domestic use are larger in volume than ice cubes made by a restaurant's or bar's ice machine. As a reasult, bartenders could not fit as much ice into a mixed drink as they could using the smaller cubes. With a smaller volume of ice per drink, more liquor and mixer had to be added, which increased costs.

Finally, restaurants and bars found it necessary to assure their patrons that precautionary measures had been taken to protect their health. Most establishments posted signs in their lobbies announcing that actions had been taken to ensure a safe water supply; others advertised in the local newspapers proclaiming safe water supplies. Perugino's Restaurant in Luzerne Borough, for example, ran a quarter-page advertisement three times a week for four weeks to assure its customers of safe water. Even so, Perugino's and several other popular restaurants and taverns in the outbreak area reported a loss of business, particularly in the first few weeks of the outbreak. Perugino's claimed to have suffered a 25 to 40 percent loss of business in the initial weeks of the outbreak. Monty's, a popular Luzerne Borough tavern, reported a 20 percent decrease from the previous year in the sale of bar drinks. Monty's was also forced to cut back the hours its employees worked to compensate for the slowdown in business, but it appears to be one of the few required to take such action.

Health care and child care facilities were also affected by the outbreak of giardiasis. Nesbitt Memorial Hospital, a 205-bed facility with 900 employees that was filled nearly to capacity, shut down all water fountains, ice machines, and soft drink dispensers and substituted canned drinks for patient use. All water used in food preparation was boiled. Water for surgical use was not affected because distilled water is used.[2]

Within eight hours of the announcement of the boil-water order, Nesbitt Hospital personnel had rented eight machines that dispense hot and cold water, and set them up in patient areas. For the ninety-nine days during which the hospital was under the boil-water order, springwater was its major source of drinking water, although a local beverage company sent the hospital more than 600 cases of canned water donated by the Anheuser-Busch Company in New Jersey. Employees living outside the affected area also brought water from home.

The experiences of the Leader Nursing Home in Kingston Borough illustrate the kinds of problems encountered by many of the businesses interviewed. With 180 patients, Leader East was the largest of three nursing homes in the affected area. For the first ten days, Leader East received, free of charge, 250 1-gallon water jugs every two to three days under an existing insurance policy with Valley Farms Dairy, a local operation. The dairy terminated this service when it turned out to be

---

[2]Personal communication from Ken Getz, administrative assistant, Nesbitt Memorial Hospital, Kingston, Pa., June 20, 1984.

too expensive. Next, on the advice of the Emergency Management Agency, a county staff person picked up 40 to 50 5-gallon containers of water for Leader East from a water tanker located in Kingston. This supply was ended after only a day or two because patients complained about the water's high chlorine and sediment levels. Leader East next turned to a nursing home in an unaffected area. However, the round trip took two hours and was judged too costly after only a few days. At its own expense, the Kingston municipal government offered to supply a 500-gallon tanker to the nursing home every day for a week. This arrangement was acceptable to Leader East, but was terminated when it proved too expensive for the local government. Leader East finally obtained water from a PG&W supply center just inside Wilkes-Barre, about a five-minute car trip from the nursing home.

Most of the day-care centers affected by the boil-water advisory immediately shut off all their drinking water sources. The director of one of the larger operations, which served about 100 children, brought in 4 to 8 gallons of water from his home every day. In addition, he purchased one 5-gallon jug of springwater each day for mixing juices. Another day-care center in Kingston Borough took the extra precaution of installing a booster heater to raise the temperature of the tap water for dishwashing. Parents from unaffected areas sent water in with their children. Toothbrushing after meals and snacks was suspended at the area's day-care centers. Water play was not allowed—a necessary step that would have been inconvenient for staff and children during the summer. Some centers used "wet-wipes" to wash hands because of children's propensity to suck water from their freshly washed hands.

Dentists were especially inconvenienced by the boil-water advisory because they all had to find an alternative to pressurized tap water for their equipment. The majority of dentists installed some variation of a bypass unit consisting of a pressurized canister or tank and a series of tubings and fittings that could be spliced into the existing system. Most bypass units were manually pressurized. It took approximately thirty seconds to pump up the pressure in the holding tank to a level sufficient to last the length of an average patient visit. The canister was filled with uncontaminated water, usually bottled springwater, and refilled every few days. Some area dentists also substituted distilled water for tap water to sterilize instruments.

This discussion highlights the more visible responses to the contaminated water supply by businesses and public service agencies in the area. Although they were not all that rare, the extra precautionary measures taken by some, such as the day-care center's purchase of a booster heater and the dentist's use of distilled water for sterilizing instruments, are not representative of the entire community. The examples presented here should not therefore be interpreted as the average business's reaction to the boil-water order. Rather, they are mentioned to illustrate the variety of problems the business community had to cope with and the range of precautionary measures that were taken.

Aftermath

An outbreak is considered stabilized once measures have been taken to ensure a safe supply of water. In the aftermath, discussions and actions turn to long-term solutions to the problem, and how the costs of the outbreak and of protection against future outbreaks are to be shared.

With the emergency over in Luzerne County, attention shifted to establishing who was to blame for the outbreak and obtaining compensation for damages. By September 1984, residents and businesses had filed thirteen giardiasis-triggered lawsuits at the federal and state levels against the water utility. The majority were class-action suits that charged PG&W with negligence in protecting the water supply.[3]

Several customers withheld payment on their water bills. In January 1984, for example, about thirty residential customers sent in partial payments, deducting amounts they felt covered the additional cost of their drinking water.

Some of PG&W's larger commercial customers also took retributive action. In January 1984 the Nesbitt Memorial Hospital decided to withhold a portion of its monthly water bill equal to the costs it was incurring for the purchase of safe drinking water. At the urging of PG&W, the hospital administrator agreed to pay the outstanding bills and to file a claim against the company for the costs incurred from the outbreak. In addition to the costs of bottled water, the hospital included in its claim the costs of flushing and cleaning all systems that were shut down during the boil-water advisory. The total claim amounted to $6,819.74 and covered the period from December 23, 1983, to March 30, 1984, when the advisory was officially lifted in that area.

---

[3]According to the PG&W's annual report (Form 10-K) to the Securities and Exchange Commission for the year ending December 31, 1989, fourteen class action and twenty-one individual lawsuits were brought against PG&W in connection with *Giardia* cysts found in samples of water taken from two distribution systems: those served by the Spring Brook Intake Reservoir and by the Elmhurst Reservoir. (The latter serves the Scranton area, which was not part of the RFF case study.) The cysts were found in the Spring Brook distribution system in January 1984 and in the Elmhurst system in March 1984. According to the PG&W report, some of the suits have been dismissed or withdrawn and others consolidated, leaving five class action and sixteen individual lawsuits pending. The report states that "the Plaintiffs seek damages, the greatest of which are in excess of $20,000 in compensatory damages and in excess of $20,000 in punitive damages, for each class member. Some of the class action suits allege that the class exceeds 100,000 people. . . . [T]he actions are still in their preliminary stages" (pp. 29–30).

According to PG&W, there is little substantive evidence that people refused to pay bills because of the contaminated water supply. During the outbreak, there were only 4 percent more accounts receivable in the boil-water area than in areas not affected by the advisory. Company officials suspect that this slightly higher than normal number of late bills occurred not in protest but because all field collectors were pulled in after the outbreak was declared, no ten-day notices were mailed to affected customers, and no three-day termination-of-service notices were sent out.

To protect the area's water supply from future contamination, the Department of Environmental Resources ordered PG&W to build a water treatment plant, including filtration, near the Spring Brook Intake Reservoir. The estimated cost of this plant was $13 million. Both the EPA and the state environmental agency agreed that installation of a standard sand filtration system was the only practical means to ensure safe drinking water from the Nesbitt and Spring Brook Intake Reservoirs.

A change in water rates (that is, the rate base against which the company can earn a return from its customers) must be approved by the state Public Utility Commission. The commission ordered PG&W to compensate its affected customers with a retroactive 25 percent reduction in water bills for the time the area was on a boil-water advisory, which cost the water company $1 million in lost revenues. As for the longer-term solution, PG&W planned to recover the investment and operating and maintenance costs of the filtration system from its customers in the form of higher future water bills spread over a number of years.

Finally, PG&W planned to undertake a long-range study of the uses of its watershed lands, to monitor beaver populations in these watersheds, and to survey the watersheds for pollution caused by third parties. As part of those plans, the company instituted a program of clear-cutting trees within fifty feet of reservoirs and major tributaries feeding them to deprive beavers of a suitable habitat. Monitoring for *Giardia* cysts was to continue on a regularly scheduled basis at all PG&W's in-service reservoirs.

REFERENCES

U.S. Bureau of the Census. 1982. *1980 Census of Population*, vol. 1, chap. B, *General Population Characteristics, Part 40, Pennsylvania*. Report PC 80-1-B40, August (Washington, D.C., Government Printing Office).

Wilkes-Barre Times Leader. 1984. "PUC Told of Water Problems." Wilkes-Barre, Pa. February 8.

Appendix 5-A

Chronology of Events

This appendix provides a chronology of events leading up to the discovery of the source of giardiasis and developments subsequent to the declaration of the boil-water advisory. The chronology highlights the roles played by the Pennsylvania Gas and Water Company (PG&W) and the major governmental agencies, including the U.S. Environmental Protection Agency (EPA), the Pennsylvania Department of Health, the Pennsylvania Department of Environmental Resources, the Luzerne County Emergency Management Agency, and the Pennsylvania Public Utility Commission.

Mid-October 1983	The number of cases of giardiasis reported to the Pennsylvania Department of Health increases above the area's "normal" level of about one case per township or borough each month.
Late November 1983	Massive influx of cases of giardiasis is reported to the Pennsylvania Department of Health. Health officials begin investigation into the source of the outbreak, initially thought to be bottled water.
December 23, 1983	On the basis of evidence from the confirmed cases, the Department of Environmental Resources and PG&W jointly issue a boil-water advisory for residents in Avoca, Duryea, Exeter, Forty Fort, Kingston, Moosic, Old Forge, Swoyersville, West Pittston, and Wyoming boroughs, Hughestown, Pittston City, Pittston Township, and a section of Jenkins Township. Boil-water orders are issued for businesses and institutions in these communities. A press briefing is held.
December 26, 1983	Pennsylvania Department of Health's district epidemiologist begins a random survey of residents in the area and eventually finds a 9 percent clinical attack rate (including a 1 percent background rate) in the area served by the Spring Brook Intake Reservoir.
December 28, 1983	The Spring Brook water distribution system is sampled.
December 29, 1983	Nine beavers trapped at Spring Brook Intake Reservoir are autopsied. Two are found to be *Giardia*-positive. A fact sheet on commonly asked questions about *Giardia* and giardiasis is issued. The Spring Brook Reservoir is sampled.

December 30, 1983   EPA takes samples of water and sediment from Spring Brook Intake Reservoir for analysis.

January 1, 1984   Seventy-three cases of giardiasis are confirmed by Pennsylvania Department of Health.

January 5, 1984   A total of 123 cases of giardiasis are confirmed. The Department of Environmental Resources blames the contamination on malfunctioning septic systems in Covington, Roaring Brook, and Spring Brook Township and on a malfunctioning sewage treatment plant in Roaring Brook, all in the Spring Brook Intake Reservoir watershed.

January 6, 1984   Pringle Borough and sections of Edwardsville Borough are added to boil-water advisory. Total number of people on boil-water advisory: approximately 75,000.

January 7, 1984   PG&W increases chlorination levels at the Spring Brook Intake from 2 parts per million (ppm) to 4 ppm.

Number of confirmed cases increases by 33, to 156.

January 8, 1984   First three of thirteen eventual lawsuits are brought against PG&W for negligence. (See footnote 3 in this chapter for more current information on the lawsuits.)

January 12, 1984   EPA announces positive test results showing *Giardia* cysts in chlorinated water from Spring Brook Intake Reservoir.

Number of confirmed giardiasis cases increases by 56, bringing the total to 212 cases.

January 13, 1984   The Department of Environmental Resources contacts school superintendents of all affected school districts and all facilities it regulates to inform them of the problem and follows up with a memorandum.

January 15, 1984   The state Public Utility Commission agrees to begin investigation into *Giardia* crisis to determine whether a rate reduction or refund should be awarded to PG&W customers on the boil-water advisory.

January 17, 1984   PG&W announces a plan to deal with the giardiasis outbreak. The plan, which must be approved by the Department of Environmental Resources, includes: (1) hiring expert consultants on the transmission of *Giardia*; (2) diverting water from two other reservoirs to Spring Brook customers; (3) increasing chlorine levels; (4) installing a 2.5-mile pipeline from Nesbitt Reservoir around Spring

Brook Intake Reservoir to supply remaining Spring Brook customers on the boil-water advisory; and (5) consideration of construction of a filtration plant at Spring Brook Intake Reservoir.

Late January 1984

PG&W, with the cooperation of various municipal officials, establishes eleven free water distribution centers for those who choose not to boil water.

February 3, 1984

A total of 330 cases of giardiasis are confirmed by Pennsylvania Department of Health.

February 13, 1984

The Department of Environmental Resources orders PG&W to: (1) divert water from Campbell's Ledge Reservoir in Duryea to PG&W customers in Duryea and parts of Pittston City; diversion should begin March 1; (2) divert water from Lake Scranton in Dunmore to the affected area; (3) divert water from Huntsville Reservoir to communities on the west side of the Susquehanna River; two months after diversion begins, people can be taken off the boil-water advisory; (4) construct the proposed pipeline around Spring Brook Intake Reservoir to uncontaminated source of water at Nesbitt Reservoir by fall 1984.

Until Nesbitt pipeline is completed, about 25,000 residents of Pittston City, Jenkins Township, Hughstown, and Moosic, Avoca, Duryea, Exeter, and West Pittston boroughs will remain on the boil-water advisory.

March 28–29, 1984

The Pennsylvania House Conservation Committee, chaired by Rep. Camille George (D–Houtzdale), holds investigative hearings in Scranton on recent outbreaks of giardiasis in the state.

March 30, 1984

EPA tests show no *Giardia* cysts in Huntsville Reservoir; residents in the boroughs of Edwardsville, Kingston, Luzerne, Pringle, Swoyersville, Forty Fort, and West Wyoming, plus some sections of Exeter Borough, are taken off the boil-water advisory.

April 9, 1984

An additional 2,303 PG&W accounts (2,163 residential accounts and 140 other accounts) are taken off boil-water advisory.

April 20, 1984

Another 59 PG&W accounts (31 residential accounts and 28 commercial accounts) are taken off boil-water advisory.

April 24–26, 1984

A State Senate committee, chaired by Senator Frank O'Connell (R–Kingston), holds public hearings on the giardiasis outbreak. Hearings focus on (1) the effectiveness of the Department of Environmental

Resources in policing water quality, (2) the quality of water treatment facilities operated by water companies, and (3) propositions for future state drinking water legislation.

May 9, 1984

Another 1,730 PG&W accounts (1,627 residential accounts and 103 other accounts) are taken off boil-water advisory.

May 26, 1984

Another 1,564 PG&W accounts (1,464 residential accounts and 100 other accounts) are taken off boil-water advisory.

July 20, 1984

The Pennsylvania Public Utility Commission orders PG&W to refund 25 percent of their water bills to Spring Brook customers who had to boil water. Residential customers are refunded an average of $10.20 for each quarter they were advised to boil water.

Eight new confirmed cases of giardiasis are reported to the district epidemiologist in Kingston, for a total of 370; residents still on the advisory are reminded to use only boiled or bottled water for drinking.

September 21, 1984

Pipeline from Nesbitt Reservoir completed; EPA tests show water to be free of *Giardia* cysts.

The boil-water advisory is lifted for the remaining 21,000 residents in the outbreak area, including residents of Pittston City, parts of Jenkins Township, Pittston Township, and Avoca, Duryea, Exeter, Hughestown, Moosic, Old Forge, and West Pittston boroughs.

PG&W publicly announces plans to keep beavers, which can carry *Giardia* cysts, out of watershed lands by clear-cutting trees within 50 feet of the banks of reservoirs and the major tributaries feeding them.

Appendix 5-B

Boil-Water Advisory

COMMONWEALTH OF PENNSYLVANIA
DEPARTMENT OF ENVIRONMENTAL RESOURCES
COMMUNITY RELATIONS
90 East Union Street—2nd Floor
Wilkes-Barre, PA 18701-3296
717/826-2511

January 17, 1984

News Desk   CONTACT: Mark Carmon
See Attached   Community Relations Coordinator

    In response to the current giardiasis outbreak, the Department of Environmental Resources (DER) has issued the following advisory to eating and drinking establishments in the affected areas:

1) All drinking water provided must come from some other approved source or be heated to a rolling boil for one (1) minute.

2) The use of filters, distillers or other purification devices is not endorsed by the Department because no units have been approved by the National Sanitation Foundation or the U.S. Environmental Protection Agency.

3) Any food prepared by adding water, but not heated to the equivalent of boiling, must be prepared using water approved for drinking purposes.

4) Fresh fruit and vegetables that will be consumed without cooking (i.e. salad bar) must be washed/prepared using water approved for drinking.

5) Ice used in drinks or in contact with food must be manufactured from water approved for drinking. Freezing the water will not kill giardia cysts.

6) All drinks (coffee, tea, soda, etc.) utilizing water must be made from water approved for drinking.

7) Water used for washing and rinsing dishes/utensils and equipment should be water approved for drinking. However, if that is not practical, all such items should be washed, rinsed, sanitized and *thoroughly air dried* before storage and/or reuse.

This advisory has been issued to protect patrons from being infected by giardiasis. Further information for proprietors of eating and drinking establishments, and the general public, is available by contacting DER's Bureau of Community Environmental Control at 826-2532 (Luzerne County) and 961-4521 (Lackawanna County).

Note: Although this announcement issued by the Pennsylvania Department of Environmental Resources is dated January 17, 1984, all affected establishments were notified of the order on December 23, 1983.

6 / Estimating Losses to Individuals

This chapter provides estimates of the losses to individuals resulting from the outbreak of giardiasis described in chapter 5. Two categories of losses are estimated—those due to illness and those due to actions taken by individuals to avoid drinking contaminated water. Estimates of losses are presented for three scenarios, which differ by the wage rates used to value loss of work and of leisure activities due to illness and by the time spent on activities, such as boiling water and obtaining bottled water, to avoid becoming ill (what is called averting or avoidance activities). A comparison of losses for the different scenarios demonstrates the sensitivity of estimates to alternative assumptions, and also the range in estimates.

LOSSES DUE TO ILLNESS

In September 1984, the Pennsylvania Department of Health mailed a questionnaire (see appendix 6-A) designed by Resources for the Future (RFF) to the 370 individuals in Luzerne County who had had confirmed cases of giardiasis. The purpose of the questionnaire was to gather data to estimate the costs incurred by those who were ill. Respondents were promised confidentiality.

By November 1, 1984, 182 victims (or their parents) had returned the questionnaire, a response rate of 49 percent. Five of the questionnaires were not used because the respondents claimed their illness began before October 1, 1983, nearly three months before the outbreak was officially declared. These individuals apparently had preexisting conditions that could not be definitively linked to the outbreak. A sixth questionnaire was not used because the respondent had a confirmed, but asymptomatic, case.[1]

Description of Survey Data

Table 6-1 provides descriptive statistics for the 176 respondents whose questionnaires were used. The average length of illness was sixty-three days, an extraordinarily high figure in comparison with other outbreaks of giardiasis. Even when the influence on this average of the prolonged illnesses of some victims was taken into consideration by using the median (rather than the average) as the measure of central tendency, the length of illness was still quite long—thirty days. Moreover, two-thirds of the sample experienced acute symptoms every day of their illness, for an average duration of illness of forty-two days. The remaining one-third experienced symptoms intermittently over an eighty-five-day period and were ill an average of sixty-two days during that period. Thus, those with intermittent symptoms were sick almost 1.5 times longer than those with continuous symptoms. This finding suggests either that intermittent illnesses were harder to treat, or that the intermittence discouraged individuals from seeking medical treatment or from following through with the course of medication prescribed, or both.

The severity of the illness can also be gauged by the fact that only 4 percent of the sample (seven people) reported being sick ten or fewer days. This finding contrasts with the results of a Department of Health (DOH) survey conducted during the outbreak, which found that 44 percent of the confirmed cases reported symptoms lasting fewer than ten days. Because it casts doubt on the credibility of both surveys, this discrepancy is troublesome and is therefore pursued here in some detail.

The discrepancy could result from one or more of four factors: (1) differences in the definition of the illness, (2) truncated responses, (3) recall bias, and (4) survey return bias.

The two surveys defined the illness differently. The DOH survey asked only for the number of days of diarrhea, while the RFF survey asked for the number of days of illness. The latter definition elicits illness

---

[1]This case was dropped because the cost per illness estimated for the confirmed cases is used later to estimate the costs of illness for those reporting giardiasis-like symptoms (the clinical cases). Including the medical cost of the asymptomatic case would lower the average cost of a symptomatic case because asymptomatics incur little or no cost.

Table 6-1.   Descriptive Statistics, Taken from Survey Results, of Confirmed Cases of Giardiasis in Luzerne County, Pennsylvania

Survey data		
Individuals surveyed (no.)[a]	370	
Questionnaires returned by November 1, 1984 (no.)	182	
Questionnaires used in the analysis (no.)	176	
Response rate (percent)	49	
Sample description (percent)		
Male	42.4	
Female	57.6	
Age distribution		
0–6 years	6.7	
7–12	2.8	
13–20	6.7	
21–30	19.2	
31–40	25.3	
41–50	16.9	
51–60	11.2	
Over 60	11.2	
Total	100.0	
Status (percent)		
Employed	58.70	
Homemaker	15.70	
Student	9.90	
Preschool	4.65	
Retired	6.40	
Unemployed	4.65	
Total	100.00	
Pretax personal income (1984 dollars per year)		
Median income group	$7,500–12,500	
Maximum income in sample	$75,000	

Disease characteristics of sample	*Average*	*Maximum*
Mean length of illness (days)	63.0	365
Median length of illness (days)	30.0	—
Respondents with continuous symptoms (percent)	65.9	—
Respondents with intermittent symptoms (percent)	34.1	—
Mean length of illness		
If continuous (days)	41.6	310
If intermittent (days)	85.0	365
Well days	23.1	—
Net sick days	61.9	300
Weight lost (pounds)[b]	9.0	34
Mean number of visits to doctor	2.0	11
Mean number of days in hospital	0.5	14
Percentage of confirmed cases first reported, by month		
October 1983	15.0	
November 1983	10.8	
December 1983	64.5	
(December 1–23)	(54.8)	
(December 23–31)	(9.7)	
January 1984	8.6	
(January 1–14)	(6.5)	
(January 15–31)	(2.1)	
February 1984	1.1	
Total	100.0	

Table 6-1 (continued)

Medication (no. of prescriptions per confirmed case)	Average	Maximum
Brand name: Flagyl (metronidazole)	1.09	3
Brand name: Atebrine (quinacrine)	0.09	3
Brand name: Furoxone (furazolidone)	0.04	2
Other	0.18	8
Economic effects[c]		
Workdays lost by:		
Employed	6.3	60
Homemakers	12.7	90
Caretakers		
Employed	0.4	—
Homemakers	12.0	—
Productivity loss (percent)[d]		
Workers	30.4	—
Homemakers	34.0	—

Note: Dash = not applicable.

[a]Questionnaires (see appendix 6-A to this chapter) were mailed on September 13, 1984, to the 370 people with confirmed cases of giardiasis in Luzerne County.

[b]The questionnaire asked respondents how many pounds they had lost as a result of their illness. Some respondents may have interpreted this question as asking for permanent weight loss, that is, net weight loss after taking into account any weight gain following their illness. Because the average weight loss of 9 pounds in the RFF survey matches closely the 10-pound average offered in the clinical literature, it would appear that few respondents interpreted the question this way.

[c]Virtually all respondents reported that their leisure activities had also been affected.

[d]These are subjective estimates of the average percentage loss in productivity for days ill but at work.

lengths equal to or greater than the former. Indeed, fatigue was the longest-lasting symptom from those respondents with long illnesses.

Truncated responses occur when illness continues after a survey is taken. Because the Department of Health survey was conducted during the outbreak, it could contain truncation bias. In fact, some illnesses were most likely continuing, because the questionnaire was mailed to an individual soon after his or her case was confirmed.

The third factor, recall bias, is the tendency to overestimate or underestimate effects due to the lag in time between the experience and the survey-induced recall of the experience. The respondents in the RFF survey had to recall the length of an illness that had ended as much as nine months before they received the questionnaire. Unfortunately, a test for recall bias is difficult to design. Assume for the moment that between the time of the illness and the receipt of the RFF survey, the length of illness became inflated in the respondents' minds. If that were the case, those who were sick earlier in the outbreak might be expected to report longer periods of illness. That hypothesis can be tested. Indeed, an analysis of the data from the RFF sample of confirmed cases showed this relationship. These results are inconclusive, however, because the same relationship would be expected to hold for an entirely different reason. The early cases, which began before the outbreak was declared, were more likely to have been misdiagnosed and thus to have received inappropriate treatment. Moreover, not realizing the long-term nature of the disease, those early cases with acute symptoms may have had little motivation to seek medical care. Conversely, after the outbreak was declared, giardiasis symptoms might be expected to be of shorter duration because treatment would be better targeted and the victims, realizing what their illness probably was, would seek immediate treatment. Because of these confounding effects, it was not possible to test definitively for recall bias.

The fourth factor, survey return bias, is the tendency of a subgroup with particular characteristics to be overrepresented in the survey sample. Those who returned the RFF questionnaire might have had longer illnesses on average than those who did not return it. If so, one might expect to find a more timely response from those with the more severe or longer-lasting symptoms. The hypothesis that the length of illness affected an individual's willingness to participate in the RFF survey can be tested to some extent by comparing the date the questionnaire was returned with the length of the illness. A regression of the number of days between the time the questionnaire was mailed and the time it was returned, on days ill, age, gender, and weight lost (used as a measure of severity) revealed significant age and gender effects. Older respondents and female respondents sent their questionnaires back earlier. However, this same regression indicated insignificant severity and "ill-days" effects. Thus, the hypothesis of survey return bias was rejected, and the survey data on length of illness were concluded to be reliable and representative of those who were ill.

The timing of the onset of confirmed cases is interesting to note (table 6-1). As expected, most cases (80.6 percent) began before the outbreak was announced in the press on December 23, 1983; 64.5 percent of all cases began in December. An additional 6.5 percent began in early January, still early enough to be within the incubation period for *Giardia* cysts ingested before the boil-water advisory was declared. Only 3.2 percent of the cases began after the middle of January 1984.

The remaining data in the table are self-explanatory. Metronidazole (brand name Flagyl) was, by far, the medication of choice in spite of its high price relative to quinacrine (brand name Atebrine).[2] Note also that average number of workdays lost was two times larger for homemakers than for those employed outside the home. This finding is not surprising. Unlike a homemaker, an employed worker forgoes either income or the opportunity to take future sick leave when he or she stays home due to illness.

---

[2]Physicians may have prescribed more Flagyl because they commonly use it to treat other diseases and are therefore more familiar with the drug and its side effects.

The relationship between workdays lost and income can be elucidated by regressing the average workdays lost on income, a dummy variable for paid sick leave, and other variables that could influence the number of workdays lost—age, gender, length of illness, and number of pounds lost (as a measure of the severity of the illness). Workers with higher incomes would be expected to lose fewer workdays for three principal reasons. First, they may feel more needed in their jobs than do workers with lower incomes, or they may like their jobs more, or both. Second, they lose more income or potential income. (This reason applies particularly to the self-employed.) Third, they were not as ill because they sought medical treatment earlier in the illness. We have no hypothesis concerning the age of victims. Those with paid sick leave would be expected to lose more workdays because they do not lose income when they stay home. Those with severe cases of giardiasis may also be more likely to stay home from work. Poisson regression, instead of ordinary least squares, was used to estimate the relationship between workdays lost and income because the dependent variable is limited to positive integers (Maddala, 1983).

Table 6-2 shows the results of the Poisson regression analysis. (The gender variable was eliminated because it was highly correlated with some of the other independent variables.) All the correlation coefficients between the remaining independent variables are low. All variables take the expected signs and are significant at the 90 percent level, and all but the number of days ill are significant at the 95 percent level. Older workers lose fewer workdays than younger workers, and those with higher incomes lose fewer workdays than those with lower incomes.

To examine whether the early medical treatment hypothesis explains the income result, we regressed income and other independent variables on the number of days ill (results not shown), reasoning that those with

Table 6-2. Explanation of Variations in Workdays Lost, Using Poisson Regression (N = 82)

Variable (X_i)	Coefficient	Asymptotic 95 percent confidence interval		Marginal changes in workdays lost[a] $\left.\dfrac{\partial WLD}{\partial x_i}\right\|_{\bar{x}_i}$
		Lower	Upper	
Intercept	4.083[b]	2.377	5.789	—
Income ($1,000 per yr)	−0.242[b]	−0.315	−0.169	−4.132
Paid sick leave (Y/N)	1.629[b]	1.053	2.206	0.795
Age	−0.0605[b]	−0.109	−0.0119	−2.333
Number of days ill	0.00277[c]	−0.00043	0.0060	0.1478
Pounds lost	0.0476[b]	0.00828	0.0869	0.5055

[a]WLD represents workdays lost.
[b]Coefficient significant at the 95 percent level.
[c]Coefficient significant at the 90 percent level.

higher incomes would seek the care of a physician earlier in their illness than those with lower incomes. If this reasoning is correct, then the number of days a person is ill should be negatively related to his or her income. However, we found that income was not related to the number of days ill and rejected the hypothesis.

Procedure Used to Estimate Losses

Based on the theory developed in chapter 3, nine categories of losses to individuals were estimated: (1) doctor visits, (2) hospital visits, (3) emergency room visits, (4) laboratory tests, (5) medication, (6) time and travel losses associated with medical treatment, (7) work loss, (8) productivity loss, and (9) loss of leisure time. Estimates were made for three alternative scenarios that differed with respect to the value of time assigned to homemakers, retirees, and the unemployed.

The costs of medical treatment and medication were obtained from local medical facilities and pharmacies, respectively. The allowable federal deduction for business automobile travel—a conservative 20.5 cents per mile—was used to estimate the cost of transportation. Time losses associated with medical treatment were valued at the appropriate wage rates.

Work time lost by those in the paid labor force was valued at the hourly wage rate before taxes for the sample of confirmed cases (on average, that rate was $8.09). The wage rate for employed persons in the sample was computed by dividing their reported before-tax annual personal income by the hours they worked that year.

Time lost by homemakers, retirees, and unemployed persons also has value, but for persons in these categories there is not even a labor-market surrogate. We used three different values of time for such persons, denoting the resulting estimates by scenarios A, B, and C. In scenario A the hourly wage rate for nonwork-time losses to groups in and out of the labor force was assumed to be $6.39, our estimate of the average after-tax wage of the respondents in the survey.[3] This estimate

---

[3]The before-tax wage rate was used to value lost work time, assuming that labor market competition acts on before-tax, or gross, wages. Using net wages would imply that worker productivity varies according to the size of employee tax exemptions. The after-tax wage was used to value nonwork-time lost, because when individuals make decisions about how to spend their nonwork time, they consider the value of that time to *them* (i.e., after taxes) in their decisions.

Unfortunately, accurate conversion of before-tax to after-tax rates requires the marginal tax rate of each household, wage rate data that were unavailable. Instead, the 1983 average tax rate on individual income (federal income tax, state and local income taxes, and FICA) of 21 percent, taken from the 1984 *Statistical Abstract* (U.S. Bureau of the Census, 1984) was used.

is not far from an estimate made by Cooper and Rice in 1976 of the hourly value of household work ($6.08 in 1984 dollars), imputed from the cost of obtaining homemaker services in the labor market. Scenario B values the time of homemakers, retirees, and unemployed persons at $2.65, the estimated minimum wage after taxes. In scenario C, the wage rate of this group is valued at zero.

To make the valuation of the time of children consistent, the value of time assigned to children 18 years of age and younger should also be based on their marginal social product. A value for the marginal social product (or wage) of working teenagers was not needed because no one in this class returned a questionnaire. For nonworking, ill teenagers and children, the value of time lost was assumed to be zero. That may not be a reasonable assumption; it was made only because there were no data for a credible estimate. Many teenagers have paid jobs and many children do chores that make valuable contributions to society.[4] To the extent that these activities took place, the assumption of a zero value of time for this age group results in an underestimation of losses. Note that time lost by parents who cared for sick children was valued at the real or implicit wage, depending on whether or not the parent was employed.

Losses in productivity were valued using the real or implicit wage rate, as applicable. Losses of leisure time were estimated by multiplying three variables: (1) the number of hours per day ordinarily available for leisure activities (the number of nonworking, nonsleeping hours), (2) the real or implicit hourly wage rate, and (3) the number of days a person was ill. For employed persons, the after-tax wage was used. This loss was reduced for those days the respondent said his (or her) productivity at work was less than 100 percent due to the illness, implying that he (or she) was not 100 percent incapacitated by the disease.

This procedure does not fully capture losses of leisure time associated with special, highly valued leisure activities such as vacations and honeymoons. The outbreak occurred just before Christmas, disrupting the holiday period for many residents. Ski trips were cancelled; an eight-day honeymoon in Hawaii was ruined by the disease. No attempt was made to value these "lost" special occasions.

Nor was a value placed on pain, suffering, stress, or anxiety, or any other psychological or physiological consequence of the outbreak. Fi-

---

[4]Children and teenagers invest in their future during these years through their physical and mental development. Depreciation of this investment through illness should, technically, be valued. That is, even if a child does not work or do chores, a serious, long-lasting illness could slow or reduce the child's development, thereby reducing his or her marginal social product in the future. Giardiasis is not regarded as a serious enough disease to induce these long-term effects, although instances of significant weight loss in infants may have long-term consequences for development.

nally, the costs of misdiagnosis were not explicitly valued, although the medical information captured some of these costs.

Estimates of Average Losses Due to Illness

Table 6-3 provides estimates of the average losses for the confirmed cases of giardiasis in the sample, using the three scenarios for valuing the wage rates of homemakers, retirees, and the unemployed. The first five items in each are estimates of the direct medical costs, paid either by the individual or by an insurance company. These are the same for all three scenarios and amount to $254 for each confirmed case. The next two items are the value of the time spent on medical treatment and services and the value of the lost work time (including housework). Adding these two items to the direct medical costs brings the estimates of losses to $631 for scenario A, in which time lost to homemakers,

Table 6-3. Average Losses for Confirmed Cases of Giardiasis in Luzerne County, Pennsylvania (1984 dollars per confirmed case)

| Loss category | Scenario[a] | | |
	A	B	C
Direct medical costs			
Doctor visits	36	36	36
Hospital visits	100	100	100
Emergency room visits	27	27	27
Laboratory fees	63	63	63
Medication	28	28	28
Subtotal	254	254	254
Time costs for medical care[b]	18	15	12
Value of workdays lost	359	271	209
Cumulative subtotal	631	540	475
Loss of productivity	371	316	278
Loss of leisure time	876	651	492
Total	1,878	1,507	1,245

[a]Scenario A assumes an implicit after-tax wage rate for the unemployed, homemakers, and retirees equal to $6.39 per hour (average after-tax wage rate of the employed); Scenario B, an implicit after-tax wage rate for the unemployed, homemakers, and retirees equal to $2.65 per hour (after-tax minimum wage); Scenario C, an implicit after-tax wage rate for the unemployed, homemakers, and retirees equal to zero. After-tax wages are the before-tax wages times 0.79. See text.

[b]Includes both the value of time spent to obtain medical care and the costs of travel.

retirees, and the unemployed is fully valued, and to $475 for scenario C, where that time is not valued at all. Adding the losses in productivity and leisure activities yields a total estimate of $1,878 for scenario A and $1,245 for scenario C.

The distribution of subtotal costs per confirmed case (not shown) has two peaks, one in the $100–200 range (14 percent), a second in the $900–1,000 range (5 percent). The latter peak occurs because of the high cost of hospitalization. An additional 4 percent of the cases incurred costs of $2,000 or more.

LOSSES DUE TO ACTIONS TAKEN TO AVOID CONTAMINATED WATER

To find out what actions individuals took to avoid contaminated water, Resources for the Future conducted fifty telephone interviews (see appendix 6-B) during September and October 1984. Households were chosen at random from the telephone book, and calls were made during the day and the evening. A telephone number was eliminated from the list after three attempts had been made to contact an adult resident. Eighty-four telephone numbers were called to obtain fifty interviews. No one refused to participate in the interview.

Information was recorded on the quantity and the number of times water was boiled, purchased, or hauled from the distribution centers or from other places. A yes or no response was obtained to questions concerning liquids that were substituted for tap water, changes in the frequency of dining out, and the use of filters or other means to remove or destroy *Giardia* cysts in the tap water. Interviewers asked about the principal strategy used to avoid contaminated water and whether this strategy was continued after the tap water was declared safe. In addition, they asked about changes in the uses of tap water, such as for brushing teeth, and whether the pressure, odor, taste, or appearance of the tap water had changed during the outbreak. Finally, respondents were asked if any family member had been sick with giardiasis-like symptoms— specifically, diarrhea lasting ten days or more—during the outbreak. This latter information was used to estimate the clinical attack rate for the outbreak area and to compare this rate with that estimated by the Department of Health from a telephone survey it conducted in January 1984.

The responses from these fifty households represent the responses of the affected population to a degree that can be described statistically. Assuming that the responses to a particular question are distributed in some known way about the mean, statistical theory can be used to establish a confidence interval around the sample mean. The mean response of the population to a given question falls within this interval with a given (say, 95) percent probability. For example, assuming the responses are distributed normally, with 54 percent of the sample reporting a permanent change in their use of tap water, the percentage of the population that would report a permanent change lies between 40 percent and 68 percent (54 percent ± 14 percent) with 95 percent probability (confidence).[5]

Description of Survey Data

Table 6-4 provides descriptive statistics for the sample that RFF surveyed by telephone. As the table shows, households chose from among a wide variety of strategies to ensure a safe drinking water supply. Nearly half (46 percent) either hauled water or boiled it, but not both. Only 2 percent of the households relied on bottled water alone. The rest used a combination of strategies. The households that hauled water obtained the largest quantity of water a week.

Undesirable changes in the quality of the tap water were widely reported, and half the households sampled said they had permanently changed their use of tap water as a result of the outbreak (either by drinking less or by switching permanently to bottled water).

Although 79 percent of the households did not change their dining habits, 15 percent ate out more often during the outbreak, and 6 percent ate out less. The increased frequency of dining out may be due to the added time and expense that cooking at home entailed. The increase contrasts with the survey data supplied by restaurants and bars, which almost uniformly reported a drop-off in business. A possible explanation for this apparent discrepancy is that households switched their patronage to establishments outside the affected area. Since the observed increase in dining out was relatively small, it could also be due to sampling error.

Procedures Used to Estimate Losses

We computed three estimates of the losses attributable to actions taken by the sample of fifty households to avoid drinking contaminated water— a lower bound, an upper bound, and a best estimate. These losses are

---

[5]The confidence interval for a characteristic variable (e.g., whether individuals in a sample have a clinical case of giardiasis) is calculated from $d = (t^2 pq/n)^{1/2}$, where t is the number of standard deviations under the normal distribution that leaves 5 percent of the area in the tails of the distribution, p is the estimated proportion of cases with the characteristic, $q = (1 - p)$, and n is the sample size. p is assumed to be normally distributed, and each of the n sampled individuals is assumed to be drawn randomly from the population (Cochran, 1963).

Table 6-4. Descriptive Statistics, Taken from Results of Telephone Survey, of Averting Behavior in Luzerne County, Pennsylvania

Statistic	Percent of households[a]	Frequency per week	Quantity (gallons per household per week)	Percent nonjoint activity[b]
Strategy for obtaining clean water				
Haul water only	22.0	1.6	10.6	36
Boil water only	24.0	2.9	6.3	32
Purchase bottled water only	2.0	1.2	5.6	16
Haul and boil water	6.0	—	—	—
Haul and purchase bottled water	18.0	—	—	—
Boil and purchase bottled water	18.0	—	—	—
Haul, boil, and purchase bottled water	8.0	—	—	—
None of the above	2.0	—	—	—
Total	100.0	—	—	—
Substitution of other liquids	53.0	—	—	—
Use of home filter	6.2	—	—	—
Permanent change in water consumption	54.0	—	—	—
Reports of undesirable changes in public water supply				
Pressure	17.0	—	—	—
Odor	73.0	—	—	—
Taste[c]	6.0	—	—	—
Appearance	58.0	—	—	—
Changes in dining out				
Increased	15.0	—	—	—
Decreased	6.0	—	—	—
No change	79.0	—	—	—
Attack rate	9.5[d]	—	—	—

Note: Dash = not available.

[a]Fifty households with a total of 148 people were contacted. The average number of people per household was 2.96.

[b]The activity was not combined with a normal everyday activity such as grocery shopping.

[c]This estimate may be low because few people drank publicly supplied water after the boil-water advisory was issued.

[d]Fourteen cases out of a sample of 148 people.

depicted in figure 6-1. This figure shows an ordinary demand curve for drinking water with an initial quantity consumed (Q_0), an outbreak-related quantity consumed (Q_1), an initial price per gallon (P_0), and the weighted average price of a substitute gallon (P_1). For lack of any better information, the demand curve for water was assumed to be linear between (P_0, Q_0) and (P_1, Q_1).

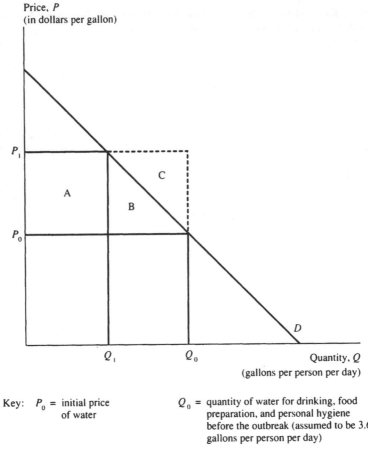

Price, P
(in dollars per gallon)

P_1

C

A

B

P_0

D

Q_1 Q_0

Quantity, Q
(gallons per person per day)

Key: P_0 = initial price
of water

Q_0 = quantity of water for drinking, food
preparation, and personal hygiene
before the outbreak (assumed to be 3.65
gallons per person per day)

P_1 = weighted average
price of substitute
water

Q_1 = quantity of water for drinking, food
preparation, and personal hygiene
during the outbreak

Figure 6-1. Demand for drinking water.

Following chapter 3, the lower-bound estimate is the cost of substitutes for tap water minus the cost of an equal amount of tap water, a loss depicted as area A in figure 6-1. This estimate uses the full cost of substitutes for tap water for drinking, food preparation, and personal hygiene, including the cost of bottled water, the time and energy costs of boiling water (by cooking fuel), and the time and travel expenses of obtaining water from other sources. If one or more of these activities was combined with an ordinary activity, say, shopping for bottled water while shopping for food—the time and travel costs associated with the averting activity were assumed to be zero. The time costs of avoiding contaminated water are included only when the activity is not combined with another activity. Finally, only the averting activities that took place during the boil-water advisory are counted as a cost of the outbreak. Any averting activity that continued after the tap water had been de clared safe is clearly a cost to society, but one that is considered an investment in lowering the risk of a future disease rather than a cost of the outbreak under consideration.

From chapter 3, the upper-bound estimate of the cost of avoiding contaminated water was calculated by valuing the quantity of tap water forgone during the outbreak at the price of a substitute gallon less the price of tap water. The actual losses are not likely to be greater than that. Figure 6-1 depicts this loss as area A + B + C. This calculation used the weighted average implicit price of the uncontaminated alternatives (boiling, bottled water, hauling). It was also assumed that each person uses 3.65 gallons of water each day for drinking, food preparation, and personal hygiene. This figure is based on information provided by Baker (1982), Solley (1980), and the Denver Water Department (no date).

The best estimate of the losses associated with the avoidance of tap water is somewhere between the lower-bound and upper-bound estimates. Following the theory of valuing individual averting behavior presented in chapter 3, the actual losses are the additional costs of substitutes for tap water consumed during the outbreak (area A) plus the losses in consumer surplus associated with the reduction in drinking water use (area B). The best estimate of losses to individuals from avoiding contaminated tap water is the sum of the two areas, or A + B.

The aggravation and inconvenience of taking averting actions was not valued, nor was the willingness to pay for tap water under pressure or with a more acceptable odor, taste, or appearance. Finally, it was assumed that the alternative water supplies were equal in quality to uncontaminated tap water.

The costs to the average household of avoiding contaminated water were estimated for three scenarios, which differ in the price of bottled water, the value of time spent on the averting activity, and whether or not the activity is assumed to be joint with an ordinary activity. These three scenarios are developed in table 6-5.

Respondents were not asked whether they combined their purchase of bottled water with other activities. Instead, two alternative prices for bottled water were used—the average delivered price of $1.17 per gallon, and the average store price of $0.66 per gallon. The delivered price may be thought of as embodying the time of obtaining bottled water as well as the cost of the water itself. Thus, the $1.17 price was applied when the purchase of bottled water was assumed to be the only activity and the $0.66 price was applied when this activity was assumed to be combined with another activity.

Table 6-5. Descriptions of Scenarios for Estimating Losses Due to Actions Taken by Individuals to Avoid Contaminated Water During the Outbreak in Luzerne County, Pennsylvania

Description	Percent of population in affected area	Scenario		
		A	B	C
Value of time (dollars per hour)				
Status				
Working adults[a]	45.6	6.39	6.39	6.39
Homemakers	22.0	6.39[a]	2.65[b]	0
Retirees and disabled	19.5	6.39[a]	2.65[b]	0
Unemployed	5.0	6.39[a]	2.65[b]	0
Students[c]				
$\geq$ 16 and working	4.4	2.65[b]	2.65[b]	2.65[b]
$\geq$ 16 and not working	3.5	0	0	0
Weighted average value of time (dollars per hour)		6.00	4.26	3.03
Boiling water is a combined activity[d]		No	Yes	Yes
Hauling water is a combined activity[d]		No	No	No
Purchase of bottled water is a combined activity[e]		No	Yes	Yes

[a]Average after-tax wage rate for working adults in the outbreak area with confirmed cases of giardiasis (before-tax wage times 0.79), in 1984 dollars.

[b]After-tax minimum wage rate (before-tax wage times 0.79), in 1984 dollars.

[c]Children younger than 16 years old are excluded from the analysis.

[d]"No" means the respondent's answer is used in the computation. "Yes" means the averting activity is assumed to be combined with an ordinary activity.

[e]Bottled water is valued at $0.66 per gallon when its purchase is assumed to be a combined activity (Yes) and at $1.17 when it is assumed not to be a combined activity (No).

Water may be boiled or hauled in combination with another activity and therefore may involve little or no time costs. For some scenarios, the respondent's answer to the joint activity question was taken as given. For other scenarios, all averting activities were assumed to have been combined with an ordinary activity.

The values of time used in the three scenarios in table 6-5 match the values of time used in the scenarios for the losses due to illness in table 6-3. The weighted average values of time for the three scenarios in table 6-5 are the sum of the values of time assigned to working adults, homemakers, retirees, the disabled, the unemployed, working students, and nonworking students weighted by the percentages of the population in each group. It is these weighted averages that were used to convert the time required for boiling or hauling water into the per gallon costs of alternative sources of water. Since these activities were assumed to take place outside of working hours, the after-tax wage is used to value the time of employed persons.

The percentages of the population in these groups, shown in table 6-5, were taken from 1980 census data or estimated from other government sources. The time of working adults was valued at the average after-tax wage rate for the sample of confirmed cases of giardiasis in Luzerne County ($6.39 per hour). The time of homemakers, retirees, the disabled, and the unemployed was valued at this wage rate, the after-tax minimum wage rate ($2.65 per hour), or zero, depending on which scenario was used. The time of working students (16 years and over) was also valued at the minimum after-tax wage rate, while the time of nonworking students (16 years and over) was valued at zero. Children and teenagers under 16 years of age were assumed to play no role in activities designed to avoid contaminated water. Hence, this segment of the population was left out of the weighting calculations in table 6-5. Because estimates of the costs of averting activities were made *by household*, this omission does not affect the calculations.

Estimates of Average Losses Due to Averting Behavior

These procedures were used to produce lower-bound, upper-bound, and best estimates of the losses associated with actions taken by individuals to avoid contaminated water in Luzerne County; those estimates are presented in table 6-6.

The lower-bound estimates of losses to the average household in the random telephone survey range from a low of $170 to a high of $776, or from $0.40 to $1.81 per person per day. The upper-bound estimates for the average household range from $800 to $2,304, or from $1.86 to $5.37 per person per day. The best estimates of losses to the average household range from $500 to $1,540 or from $1.13 to $3.59 per person per day.

The lower-bound estimates do not consider certain types of averting behavior, such as the substitution of other liquids for water (the purchase of premixed orange juice, for example, instead of frozen) that would increase costs. Whether the best and upper-bound estimates include these costs depends on assumptions made about the price of these substitutes relative to water.

Caution should be taken when interpreting the data presented in table 6-6. They are subject to a number of potential biases as well as other shortcomings. First, the selection of households to contact by telephone may not have been random. Ideally, calls should be made at times that would give each household an equal probability of being contacted. In

Table 6-6. Average Costs of Actions Taken by Households to Avoid Contaminated Water During the Giardiasis Outbreak in Luzerne County, Pennsylvania (1984 dollars)

	Scenario		
Average costs	A	B	C
Lower-bound estimate[a]			
Total cost per household	776.40	182.80	169.60
Cost per person per day	1.81	0.43	0.40
Upper-bound estimate[b]			
Total cost per household	2,304.40	852.40	800.00
Cost per person per day	5.37	1.99	1.86
Best estimate[c]			
Total cost per household	1,540.40	517.60	484.80
Cost per person per day	3.59	1.21	1.13

[a]The lower-bound estimate is based on the actual cost of the substitute water used for drinking, food preparation, and personal hygiene.

[b]The upper-bound estimate is based on the cost of replacing with a substitute source of water the quantity of water for drinking, food preparation, and personal hygiene that would have been used had the municipal water supply not been contaminated. These are not actual costs, but short-run opportunity costs.

[c]The best estimate is the lower-bound estimate plus one-half the difference between the upper-bound and lower-bound estimates. (See text for fuller explanation.)

particular, if telephone calls are made only during the day, households with all working adults are unlikely to be contacted. Excluding working adults could have important consequences for valuing averting behavior if the mitigating strategies followed by households with all adults working differed from those in the survey. Although a special effort was made to reach households in the evening when most working adults would be home, the sampling procedure may be biased toward households with nonworking adults. Fifty-nine percent of the households in the sample had nonworking adults; nationally, fewer than 50 percent of all households have nonworking adults.

Even if such a bias existed in the sampling procedure, the estimates of losses due to averting behavior would be biased only if the different types of households had adopted strategies with different costs. This hypothesis was tested by including in regressions that explained variations in costs a variable for whether the household included a nonworking adult. This variable was not significant.

Another potential problem concerns the characterization of averting behavior patterns. Our interest in limiting the time of the telephone interviews and the recall difficulties respondents would have faced had they been asked to explain their behavior in minute detail argued for simplification of the survey questions. As a result, respondents were asked to describe only the mix of averting activities they took immediately after the outbreak had been declared and the mix they finally settled upon. Interim adjustments in averting activities were ignored.

A related issue is the treatment of averting behavior that was combined with other activities. If a respondent said that an averting activity was combined with an ordinary activity, we assumed it was combined throughout the period of the boil-water advisory. Undoubtedly, the situation was more complex than that. The existence and direction of bias introduced by this assumption are not known because activities classified as combined may sometimes not have been, and vice versa. This issue is addressed in the analysis by assuming in scenarios B and C that boiling water is always a combined activity and assuming in scenario A that the reported response to the question whether boiling water was a combined activity is an accurate characterization of actual behavior (see table 6-5).

Households with one or more family members with giardiasis did not incur different averting costs than households free of the illness. Nor did those households that made permanent changes in their use of tap water follow different averting behavior strategies.

A final problem concerns the use of a single demand curve for water and the assumption that it is linear over part of its range. The concept is an abstraction that ignores the different attributes individuals assign to water depending on its source: tap, bottled, hauled, and boiled. Bottled water, for instance, may be regarded as having superior taste and color to tap water. Only if these four sources of water are seen as perfect substitutes can a single demand curve for water be said to exist. Otherwise, water from each source has its own demand curve, and these curves may be related to one another in complex ways. The assumption of a linear demand curve may bias the best estimate up or down depending on its true shape. In the absence of information on the actual shape of the demand curve, the linear assumption seems as good as any other.

TOTAL LOSSES TO INDIVIDUALS IN THE OUTBREAK AREA

Having determined a range of estimates of average losses for the 176 people with confirmed cases of giardiasis who returned the questionnaire and for the fifty households that responded to questions about the actions they took to avoid drinking the contaminated water, the final step was to estimate the total losses to the 75,000 people at risk in the outbreak area.

The procedure used to estimate total losses due to actions taken to avoid contaminated water is straightforward. It assumes that the sample of fifty households is representative of the 25,000 households in the affected area and involves multiplying the average total cost per household (table 6-6) by the 25,000 households at risk (table 5-2). Following this procedure, total lower-bound estimates of the losses due to averting activities for scenarios A, B, and C in table 6-7 range from $4.2 million to $19.4 million; total upper-bound estimates of the losses range from $20.0 million to $57.6 million. The best estimates are midpoints, which range from $12.1 million to $38.5 million.

To estimate the total losses in the affected area due to illness, the number of persons who were ill with giardiasis had to be estimated. The estimate we used—6,000—was determined by multiplying the Department of Health's estimate of an 8 percent net attack rate (over and above the background rate of 1 percent) by the 75,000 people at risk (table 5-1).[6] Of the 6,000, 370 people had confirmed cases of giardiasis; the remaining 5,630 people were clinical cases.

The results of the RFF random telephone survey produced a net attack rate of 8.5 percent. To err on the conservative side, the slightly smaller Department of Health rate was used to calculate losses.[7]

The 8 percent attack rate is a point estimate based on a sample, and the actual rate may be greater or less than this estimate. Given that 233 people living in the outbreak area were sampled in the Department of Health survey and that the sample net attack rate is 8 percent, there is a 95 percent probability that the true net attack rate is between 4.5 percent and 11.5 percent of the exposed population (the sample mean of 8.0 percent ± 3.5 percent).[8]

The procedure used to scale average losses calculated for the sample to the appropriate population is on potentially shaky ground for estimating losses due to illness. The survey of the confirmed cases of giardiasis may not be representative of those who were ill with giardiasis. If the group surveyed differs from others who were ill in ways likely to influence responses to illness, a simple proportional scaling of average

[6]Letter to J. Glenn Gooch, president, Pennsylvania Gas and Water Company, from Thomas M. DeMelfi, epidemiology investigator, Pennsylvania Department of Health, May 23, 1984.

[7]A confidence interval can also be computed for the RFF estimate of the net attack rate. Because only 148 people in fifty households were sampled, the confidence interval around the RFF point estimate is wider than that around the health agency's estimate. The RFF confidence interval is the sample mean (8.5 percent) ± 4.5 percent, or from 4 percent to 13 percent. This interval calculation assumes independence of the sample observations. Assuming dependence within families widens this interval to the sample mean ± 7.7 percent, or from 0.8 percent to 16.2 percent.

[8]Since some of these individuals may be members of the same family and since giardiasis is known to strike entire families, the n observations may not be independent. Assuming

Table 6-7. Total Losses Due to Illness and to Actions Taken by Individuals to Avoid Contaminated Water During the Giardiasis Outbreak in Luzerne County, Pennsylvania (millions of 1984 dollars)

Loss category	Scenario[a]		
	A	B	C
Losses due to illness[b]			
Medical costs	1.07	1.05	1.03
Loss of work	2.15	1.63	1.25
Loss of productivity and leisure time	7.48	5.80	4.62
Total losses due to illness	10.70	8.48	6.90
Losses due to averting actions[c]			
Lower-bound estimate	19.41	4.57	4.24
Upper-bound estimate	57.61	21.31	20.00
Best estimate	38.51	12.94	12.12
Total losses			
Lower-bound estimate	30.11	13.05	11.14
Upper-bound estimate	68.31	29.79	26.90
Best estimate	49.21	21.42	19.02

[a]The scenarios are described in tables 6-3 and 6-5.

[b]Estimated by multiplying average costs per confirmed case minus hospital costs by the 5,630 clinical cases (6,000 total cases − 370 confirmed cases = 5,630 clinical cases) and adding to this product the total costs of the 370 confirmed cases.

[c]The lower-bound estimate is based on the actual costs to households of providing uncontaminated water for drinking, food preparation, and personal hygiene. These costs include both out-of-pocket expenses and some time costs (see table 6-5). The losses due to individual averting actions are not likely to be less than this lower-bound. The upper-bound estimate is based on the opportunity costs of the tap water use forgone and is a maximum estimate. The losses are not likely to be greater than this upper-bound. The quantity of uncontaminated water actually used is incorporated in the lower-bound estimate. The quantity of water normally used for drinking, cooking, and dishwashing is used to estimate the upper bound. The actual opportunity cost of tap water use forgone is likely to be between the upper and lower bounds and is called the best estimate. The best estimate in this table is assumed to be the average of the upper and lower bounds and is computed as the lower-bound estimate plus half the difference between the upper-bound and lower-bound estimates. (See the text of this chapter and of chapter 3 for more complete discussion.)

complete dependence among family members and three members in each family, the sample of independent observations shrinks to seventy-eight. Then, the confidence interval is the sample mean ± 6 percent, or from 2 percent to 14 percent, which is wider than the confidence interval when independence is assumed (the sample mean ± 3.5 percent).

losses would not be appropriate. For example, if those with high incomes tend to spend more on medical treatment for giardiasis than those with low incomes, and if they are overrepresented in the sample simply because they were more likely to seek medical care, a direct scaling of average losses would overestimate total medical costs in the affected area. In this case, the appropriate approach is to determine income differences in estimates of losses, based on the sample, and use this information to assign losses by income class to the people who were ill.

A comparison of the proportion of people in the sample by age and income with the proportion of the exposed population by age and income, using data from the U.S. Bureau of the Census (1982), found no statistically significant differences. Therefore, assuming that the personal and family characteristics of the clinical cases were not statistically different from those of the exposed population, the confirmed-case sample was judged to be representative of all of those who were ill (the confirmed plus the clinical cases) with respect to age and income.

In estimating losses due to illness, every clinical case was assumed to have been giardiasis. This assumption may bias results upward because some clinical cases very likely were not giardiasis at all but another illness with similar symptoms. This assumption is examined more closely in chapter 8.

We are also concerned that the average loss per confirmed case (table 6-3) is not representative of the 5,630 people who had clinical cases of the disease. An average clinical case may be relatively less severe than an average confirmed case. An illness is confirmed only when a person is sick enough to seek medical attention and willing to provide one, two, or perhaps even three, stool samples. Thus, an illness may be unconfirmed simply because a person does not feel sick enough to visit a physician and to submit a stool sample for laboratory tests. On the other hand, some physicians apparently prescribe medication to patients with giardiasis-like symptoms without taking stool samples, particularly after an outbreak has been declared. Thus, a clinical case may be no less severe than a confirmed case, although it may be less costly. This issue is addressed more fully in chapter 8, where the results are discussed and interpreted.

In spite of these caveats, an average clinical case was judged to be just as severe and just as costly as an average confirmed case, with one exception: anyone sick enough to be hospitalized for giardiasis-like symptoms presumably would have been tested for *Giardia* cysts until a definite diagnosis was obtained. Therefore, hospital costs incurred by the confirmed cases were subtracted before scaling losses to the clinical cases and added back after losses to the clinical cases were estimated. Based on these data and procedures, the total losses due to illness range from $6.9 million to $10.7 million, as summarized in table 6-7.

REFERENCES

Baker, Larry K. 1982. "Experiences and Benefits of the Application of Minimum Flow Water Conservation Hardware," p. 282 in *Proceedings of the National Water Conservation Conference on Publicly Supplied Potable Water*, April 14–15, 1981, Denver, Colo.

Cochran, William G. 1963. *Sampling Techniques* (New York, John Wiley & Sons, Inc.).

Cooper, B. S., and D. P. Rice. 1976. "The Economic Cost of Illness Revisited," *Social Security Bulletin* (February) pp. 21–36.

Maddala, G. S. 1983. *Limited Dependent and Qualitative Variables in Econometrics* (New York, Cambridge University Press).

Denver Water Department. No date. "Hogwash," Denver, Colo.

Solley, W. 1980. "Water Use in the United States," *Popular Publications of the U.S. Geological Survey*, No. 1001 (Alexandria, Virginia, U.S. Geological Survey).

U.S. Bureau of the Census. 1982. *1980 Census of the Population*, vol. 1, chap. B, *General Population Characteristics, Part 40, Pennsylvania*. Report PC 80-1-B40, August (Washington, D.C., Government Printing Office).

U.S. Bureau of the Census. 1984. *The Statistical Abstract of the United States* (Washington, D.C., Government Printing Office).

Appendix 6-A

Questionnaire Designed by Resources for the Future and Sent by the Pennsylvania Department of Health in September 1984 to 370 Individuals with Confirmed Cases of Giardiasis in Luzerne County, Pennsylvania

This questionnaire is to be filled out by the giardiasis victim named on the envelope or, if the victim is a child, by a parent answering questions for the child.

I. *Personal and Employment Information of Ill Person*

1. Sex   1 <u>male</u>   2 <u>female</u>   (circle appropriate answer)

2. Age ＿＿＿＿＿ years

3. Zip code ＿＿＿＿＿; First three digits of telephone number (not including area code) ＿＿＿＿

4. County ＿＿＿＿＿＿＿＿＿＿＿＿＿＿＿＿＿＿＿＿

5. Did you have health insurance during your bout with giardiasis?
   1 <u>yes</u>,   2 <u>no</u>   (circle one)

6a. At the onset of your illness, were you   1 <u>employed</u>, 2 <u>retired</u>,   3 <u>a student</u>,   4 <u>housewife (husband)</u>, 5 <u>unemployed</u>,   6 <u>pre-school</u>,   7 <u>other</u>   (specify) ＿＿＿＿＿＿ (circle applicable responses)

   b. At that time did you work   1 <u>full-time</u>,   2 <u>part-time</u>, 3 <u>I did not work</u>   (circle one)

   c. If you worked, how many days a week did you work? ＿＿＿ days. How many hours a day did you work? ＿＿＿ hours

   d. Were you self-employed during this time?   1 <u>yes</u>, 2 <u>no, I worked for someone</u>,   3 <u>I did not work</u>

Only answer questions 7, 8, & 9 if you are employed.

7. Paid sick leave is an employee benefit whereby an employee continues to receive pay when he or she misses work because of sickness.

The number of paid sick days per year is usually limited. For example, the Federal government provides 13 paid sick days per year. Self-employed people do not have paid sick leave. Does your employer provide paid sick leave?   1 yes,   2 no, 3 not sure   (circle one)

8. If you answered yes to question 7, how many days per year of paid sick leave does your employer provide? _____ days

9. Do you receive paid vacation days? Self-employed people do not receive paid vacation days.   1 yes,   2 no

II. *Illness Characteristics*

10. How many days did your illness last? Count from the date you first experienced symptoms of giardiasis until the date the last symptom went away. (If it is still continuing, write "continuing.") _____ days

11. For some people, their giardiasis seems to clear up for a while and then return. Has this happened to you?   1 yes,   2 no (circle one)

12. If you answered yes, how many days were you feeling well during the entire time you had this illness? _____ days

13. Please fill in for each symptom your best estimate of the date it began and its duration. If the symptom was intermittent, give the total number of days the symptom was experienced.

	Date Began	How Long*
Diarrhea	_____ /	_____
Nausea	_____ /	_____
Vomiting	_____ /	_____

*If still continuing, write the number of days from the beginning of symptom and "continuing."

Stomach or intestinal pain   _____ / _____

Excessive gas   _____ / _____

Fatigue   _____ / _____

Other (describe)   _____ / _____

_____

14.   What is today's date?   _____

15.   About how many pounds, if any, have you lost as a result of this illness?   _____ pounds.

III. *Medical Care*

Some people may have visited a doctor, hospital, or clinic, etc., to have their illness diagnosed and to receive treatment. If you did not use any professional medical services, write in zeros for questions below.

16.   As a result of your illness, how many visits in total have you made for professional medical services?   _____ visits. (Don't count trips in which you simply dropped off a stool sample or picked up test results.)

17.   As a result of your illness, how many of the above visits did you make to:
a) The doctor's office?   _____ visits
b) Hospital emergency room?   _____ visits
c) Other (specify)?   _____ visits to _____.

18.   What is the one-way distance (miles) and travel time (minutes) to your
a) Physician?   _____ miles;   _____ minutes
b) Hospital?   _____ miles;   _____ minutes
c) Other? (specify)   _____ miles;   _____ minutes to
_____.

19.   Were you admitted to a hospital?   1 yes,   2 no   (circle one)
If yes, how long was your stay?   _____ days.

20.   Did you provide a stool sample?   1 yes,   2 no   (circle one)
On how many separate occasions did you provide a stool sample?
_____ occasions

21.   Were *Giardia* cysts found in any of these samples?   1 yes,
2 no   (circle one)

22. Was medication prescribed for your illness?   1 <u>yes</u>,   2 <u>no</u>
If yes, indicate below which medication and the number of pre-scriptions filled.

Number of
Prescriptions

Atabrine   1 <u>yes</u>, 2 <u>no</u>   _____

Quinacrine   1 <u>yes</u>, 2 <u>no</u>   _____

Flagyl   1 <u>yes</u>, 2 <u>no</u>   _____

Metronidazole 1 <u>yes</u>, 2 <u>no</u>   _____

Furoxone   1 <u>yes</u>, 2 <u>no</u>   _____

Furazolidone   1 <u>yes</u>, 2 <u>no</u>   _____

Other (specify) 1 <u>yes</u>, 2 <u>no</u>   _____

23. List any non-prescription drugs purchased for this illness, such as Kaopectate, Emetrol, etc.   _____

24. Have any of the drugs you took for the illness produced side effects?   1 <u>yes</u>,   2 <u>no</u>,   3 <u>not sure</u>   (circle one)

If yes, which drugs? What side effects? How long did they last?

		Duration of
Drug	*Side Effects*	*Side Effects*
1. _____	_____	_____
2. _____	_____	_____

25. What percent of your hospital, clinic, and physician costs were covered by health insurance? _____ percent. If you are unsure, please give your best estimate.

26. What percent of your medication costs were covered by health insurance? _____ percent

GO TO SECTION VI IF YOU ARE A PARENT ANSWERING A QUESTIONNAIRE FOR A CHILD.

IV. *Restricted Activity of Ill Person*

27. Did this illness cause you to miss work?   1 yes,   2 no,
3 I don't work   (circle one)
If so, how many days? _____ days

28. If you went to work during this illness, did your illness affect your ability to work as hard as you usually do at your job?   1 yes, 2 no   (circle one)

29. If yes, was your ability slowed: 1 a little (0–5%),   2 a fair amount (6–25%),   3 a lot (26–50%),   4 more than 50% (circle one)

30. Did this illness cause you to miss school?   1 yes,   2 no, 3 I don't go to school   (circle one)
If so, how many days?   _____ days

31. Did this illness cause you to refuse offers or cancel any plans that you had to take part in a leisure activity (movies, sports, long walks, dining out, etc.)   1 yes,   2 no   (circle one)

    List the activities you cancelled or for which you refused offers to participate. _____
    _____
    _____

32. In general, did this illness force you to change your normal routine? Do not count changes in your routine resulting from your attempts to avoid being exposed to *Giardia* cysts (such as buying bottled water).   1 yes,   2 no   (circle one)

    List the major changes forced on you by your illness.
    _____
    _____
    _____

    IF YOU ARE A PARENT ANSWERING FOR A CHILD, SKIP TO SECTION VI.

V. *Restricted Activity of Other Family Members*

33. Did anyone in your family stay home from their job to care for you (for example, for home care, or to take you to the doctor)? 1 yes,   2 no   (circle one)

    If yes, how much time, in total, did the other person(s) lose from work? _____ hours

What is the occupation of the family member(s) who stayed home from work? _____
Did the other family member take annual leave for this?
1 yes,   2 no
Did this person lose pay?   1 yes,   2 no

34. Did your illness cause other family members to refuse offers or cancel any plans that they had to take part in a leisure activity (movies, sports, long walks, dining out, etc.)?   1 yes,   2 no (circle one)

List the activities they cancelled or for which they refused offers to participate. _____
_____
_____

35. In general, did your illness force other family members to change their normal routine? Do not count changes in their routine resulting from their attempts to avoid being exposed to *Giardia* cysts (such as buying bottled water).   1 yes,   2 no   (circle one)

List the major changes forced on your family by your illness.
_____
_____
_____

VI. *For Parents Answering Questionnaires for a Child*

36. Did you or another member of your family stay home from your (their) job to care for the sick child (for example, for home care, or to take a child to the doctor)?   1 yes,   2 no (circle one)

If yes, how much time, in total, did you (they) lose from work?
_____ hours
What is the occupation of the family member(s) who stayed home from work? _____
Did you take annual leave for this?   1 yes,   2 no
Did you lose pay?   1 yes,   2 no

37. Did you incur any extra expenses for baby-sitting or day-care because of your child's illness?   1 yes,   2 no   (circle one)

If so, how much did this cost you? $ _____

38. Did your child's illness cause you to refuse offers or cancel any plans that you had to take part in a leisure activity (movies, sports, long walks, dining out, etc.)?   1 yes,   2 no   (circle one)

List the activities you cancelled or for which you refused offers to participate. _____

_____

_____

39. In general, did your child's illness force you to change your normal routine? Do not count changes in your routine resulting from your attempts to avoid being exposed to *Giardia* cysts (such as buying bottled water). 1 yes, 2 no (circle one)

List the major changes forced on you by your child's illness.

_____

_____

_____

40. Did this illness cause your child to miss school? 1 yes, 2 no, 3 My child doesn't go to school (circle one)

If so, how many days? _____ days

VII. *Income*

To further our understanding of the costs of a giardiasis outbreak, it is important for us to ask for information on your family and personal income. All responses will be held strictly confidential.

41. What was your family's gross annual income last year (before deductions for taxes)? (Your family income is the sum of the earnings from all sources for all family members living in the same household.) (circle one)

1	0
2	$1– $5,000
3	5,001– 10,000
4	10,001– 15,000
5	15,001– 20,000
6	20,001– 30,000
7	30,001– 45,000
8	45,001– 60,000
9	60,001– 80,000
10	80,001–100,000
11	100,001 or greater

42. Thinking now only of your (the ill person's) contribution to family income, what was your (the ill person's) annual earned income before taxes last year? (circle one)

1	0
2	$1–$5,000
3	5,001–10,000

4	10,001–15,000
5	15,001–20,000
6	20,001 25,000
7	25,001–30,000
8	30,001–35,000
9	35,001–40,000
10	40,001–45,000
11	45,001–50,000
12	50,001–75,000
13	75,001 or greater

43. Typically, how many weeks a year do you (the ill person) work?
    _____ weeks

Appendix 6-B

Questionnaire on Averting Behavior Used in Telephone Interview Conducted by Resources for the Future with Random Sample of Fifty Households in Luzerne County, Pennsylvania, September and October 1984

NAME: _____   FAMILY SIZE _____

ADDRESS: _____   AGE/SEX/DIARRHEA 10 DAYS
                                                      OR MORE

_____

PHONE #: _____   ___ ___ ___

AWARE OF GIARDIASIS   ___ ___ ___

    Yes   No   ___ ___ ___

WHEN AWARE _____   ___ ___ ___

TAP WATER   WHAT DID YOU STOP
QUALITY/QUANTITY CHANGE   USING TAP WATER FOR:

Pressure _____   Drinking _____
Odor _____   Cooking _____
Taste _____   Teethbrushing _____
Appearance _____   Bathing (Kids?) _____
When _____   Clotheswashing _____
How Long _____   Dishwashing _____
                                   Other _____

                                   WHEN STOP _____

CHANGE IN WATER
CONSUMPTION: COMMENTS:

None _____   _____
Very little (0–10%) _____
Somewhat (11–25%) _____   _____

A lot (26–50%) _____
Very much (50% +) _____   _____

_____

SUBSTITUTES

	How Often	How Much per Time	Date Stopped	How Far	Joint Activity/Trip	Other

Haul water[a] _____

Boil water[a] _____

Bottled[a] _____

Other liquid[a] (Yes/No; describe) _____

Change in dining out frequency (Yes/No; More/Less) _____

Filters used (Yes/No) _____

[a]Indicate most important with an "*". Write C if continuing beyond lifting of advisory.

_____

PERMANENT CHANGE IN USE PATTERN:
(Yes/No; More/Less Tap Water)

7 / Estimating Losses to Businesses and Government Agencies

Businesses, schools, government agencies, and the water utility, as well as individuals, suffered losses during the outbreak of giardiasis in Luzerne County. This chapter estimates the value of the losses for commercial and public enterprises.

The range of losses was most diverse among businesses in the affected area. Some experienced great inconvenience and costs, others virtually no inconvenience at all. Only businesses with measurable losses are included in this analysis—and even then, not all of them. Restaurants and bars, hospitals, nursing homes, dentists, and day-care centers are analyzed; meat-packers, hotels, and motels are not. Restaurants and bars are singled out for special attention because so many, an estimated 250, were affected, and because computing their losses is relatively complex.

Compliance with the boil-water order resulted in out-of-pocket expenses to private and public schools in the three affected school districts. Losses to government agencies included the value of time and the out-of-pocket expenses incurred by government officials who had key responsibilities during the outbreak.

Alternative scenarios were developed for estimating the losses to restaurants and bars, but only one was developed for other businesses, schools, government agencies, and the water supply utility. However, because the estimates of some categories of losses involve substantial uncertainty, they are separated from the others and included in table 7-5 under the heading "Possible costs." Little confidence is placed in the magnitude of these estimates, although losses were undoubtedly incurred in these categories.

LOSSES TO RESTAURANTS AND BARS

All of the approximately 250 restaurants and bars located in the outbreak area took at least some measures to comply with the boil-water order. Most also claimed to experience a falloff of sales during the outbreak. This reduction adds to the loss in social welfare through a loss in profits and a loss in consumer surplus.

In this section, the results of the survey of restaurants and bars are discussed and the procedure used to estimate losses outlined. Average losses in profits and average losses associated with actions taken to avoid contaminated water are presented for the sample, for three scenarios. These average losses are then scaled to the number of restaurants and bars in the affected area to obtain estimates of total losses.

In the absence of contrary information, it is assumed that consumers neither switched to establishments outside the affected area during the outbreak nor increased their use of establishments inside the affected area over "normal" levels after the outbreak ended. Losses in consumer surplus are calculated for affected restaurants and bars using two assumptions about the price elasticities of demand. A final section summarizes the estimates of losses for restaurants and bars and discusses the limitations of these estimates.

Survey of Restaurants and Bars

A questionnaire was sent to 250 restaurants and bars in the affected area to gather basic information about the characteristics of the establishments (number of employees, number of customers per week, and so forth), the costs of the outbreak, and market information (the percent change in prices and sales as a result of the boil-water order). This questionnaire is included in appendix 7-A. The names and addresses of affected restaurants and bars were obtained from the Pennsylvania Department of Environmental Resources. Data on profits and reductions in profits were not requested because of the confidential nature of this information. Seventeen restaurants and five bars returned the questionnaires, representing a response rate of 9 percent. Practical considerations ruled out a follow-up survey to either increase the response rate or test for the presence of response bias. The low response rate means that the estimates of average losses to restaurants and bars may be quite far from actual average costs for the 250 establishments affected.

Despite the low response rate, a wide range in sizes of establishments was represented in the sample. The number of customers per week ranged from 40 to 6,500, with an average of 1,719. The number of employees per establishment ranged from two to seventy-five. The average was fourteen.

All twenty-two establishments took steps to comply with the boil-water order. The most popular strategy, used by six of the establishments, was the most complex—a combination of bottled, boiled, and hauled water. All possible combinations of these three alternatives were represented in the sample.

Ice was also a problem. Twenty of the twenty-two establishments purchased ice. Thirteen also substituted more expensive bottled soda and mixers for those bulk drinks that had to be mixed with water. The survey revealed that only one restaurant changed its prices during the period of the boil-water order, but most establishments experienced a reduction in sales—the average decline was 10.5 percent. Most also reported that sales rebounded to previous or near previous levels after the outbreak was over. On average, sales rebounded 7.8 percent. Because the restaurant business fell off nationwide during the period immediately after the advisory was lifted (the spring and summer months of 1984), it was assumed that sales returned to their "normal" level. Given that assumption, social losses associated with the temporary reduction in the supply and demand for restaurant and bar services in the affected area were not estimated beyond the end of the boil-water order.

Procedure Used to Estimate Losses

The methods used to estimate losses to restaurants and bars follow those presented in chapter 4. The establishments were assumed to have experienced rising marginal costs, and the increase in production costs and the fall in sales were attributed wholly to the outbreak. These assumptions, coupled with the survey finding that prices were generally held constant, imply that the situation depicted in figure 4-4(b) is the most appropriate economic model for calculating social losses. In this model of markets, the demand and supply curves both shift, but prices remain constant and quantities decrease. This model identifies three components of the social losses due to the outbreak: the direct out-of-pocket costs imposed on restaurants and bars, a loss in profits because of reduced sales attributed to a reduced demand for services, and a loss in consumer surplus. The first two components constitute the loss in producer surplus.

For the sole establishment that reported a price increase, the situation depicted in figure 4-3(c) is applicable; in this scenario the supply curve shifts upward, but the demand curve is unchanged. Although, in principle, demand could have shifted too, the assumption that only supply changed yields a more conservative estimate of losses in consumer surplus.

Finally, responses to the questionnaire showed that layoffs were rare during the outbreak and that even when productivity fell, neither prices nor wages were affected. Thus, the estimates of losses ignored the effects of layoffs and of reduced productivity on producer surplus.

Direct Costs   The procedures used to estimate the costs to restaurants and bars of avoiding contaminated water are the same as those used to estimate the costs to individuals. The amount of water purchased, boiled, or hauled, the time period over which these activities occurred, and the cost of each alternative per period were estimated for individual businesses and then averaged over all reported businesses. Similar calculations were made for the purchase of ice and premixed or bottled mixers and soda used as substitutes for drinks mixed on the premises. Information on equipment purchased to cope with the outbreak is reported in table 7-1 but was not used in the analysis because the sample was too small to scale these costs with confidence to all restaurants and bars in the outbreak area.

Three scenarios are constructed. Scenario A assumes the minimum wage rate of $3.55 per hour and that boiling and hauling water are not combined with other activities. Scenario B assumes the same wage rate and that boiling water is combined with another activity but that hauling water is not. Scenario C assumes a low-skill wage rate of $5.00 per hour and that neither boiling nor hauling water is combined with other activities. Scenarios that both include and exclude bars were constructed because baseline profit data on bars were unavailable. Finally, a 70 percent price differential between case equivalents of soda and mixers was assumed.[1]

Lost Profits   To calculate lost profits for restaurants and bars (the second element in the measurement of social losses), a 1983 industry survey was obtained from the National Restaurant Association (1984), which provided average turnover rates, sales, costs, and median profit data for restaurants in the northeastern United States. Because profit

---

[1]This price differential was supplied by an official of Carbo-Mix Dispensers, Beltsville, Md., February 20, 1985.

Table 7-1. Average Costs to Restaurants and Bars in the Outbreak Area

	Scenario[a]							
	A			B			C
Loss category	Restaurants	Bars	Combined	Restaurants	Bars	Combined	Combined
Number of establishments	17	5	22	17	5	22	22
Average number of customers served[b]	41,833	4,430	32,482	41,833	4,430	32,482	32,482
Average losses:							
Due to averting activities (1984 dollars per establishment)							
Bottled water	217	262	227	217	262	27	239
Boiled water	1,259	1,151	1,233	71	57	68	1,807
Hauled water	106	36	90	106	36	90	118
Ice	912	566	834	912	566	834	834
Soda	1,230	385	1,008	1,230	385	1,008	1,008
Subtotal	3,724	2,400	3,392	2,536	1,306	2,227	4,006
Per customer	0.0890	0.5418	0.1044	0.0606	0.2948	0.0686	0.1233
Due to lost profits (1984 dollars per establishment)	3,242	N.A.	N.A.	3,242	N.A.	N.A.	N.A.
Total losses	6,966	N.A.	N.A.	5,778	N.A.	N.A.	N.A.
Total losses per customer	0.1665	N.A.	N.A.	0.1381	N.A.	N.A.	N.A.
Equipment costs (1984 dollars per establishment)	304	80	251	304	80	251	251

Note: N.A. = data were not available.

[a]Scenario A assumes a wage of $3.55 per hour and that boiling and hauling water are not combined with other ordinary activities. Scenario B assumes a wage of $3.55 per hour and that boiling water is combined with another activity. Scenario C assumes a wage of $5.00 per hour and that boiling and hauling water are not combined with other activities.

[b]Total during the outbreak period.

rates varied widely between restaurants that sold only food and those that sold both food and alcoholic beverages, the sample was divided into these two categories (six of the seventeen restaurants sold only food), and reductions in profits were calculated for each group.

Data on turnover rates were used to convert the profit per seat to profit per customer per day. Lost profits per establishment were computed as the daily profit per customer multiplied by the number of customers multiplied by the number of days the boil-water order was in effect multiplied by the proportion of output not produced during the boil-water order period (obtained from the survey questionnaire).

Because we were interested in how profits responded to variable cost changes, we used the daily profit rates before the costs of depreciation, rent, interest, and other fixed costs were subtracted from revenues. These costs would have had to be paid irrespective of changes in demand or supply induced by the outbreak.

Loss in Consumer Surplus The third element of losses to restaurants and bars—the loss in consumer surplus—was calculated using two different price elasticities of demand, -2.2 and -7.8, taken from *Food Service Trends*, published by the National Restaurant Association (1980, 1981). Because the demand curve for restaurant services (labeled D in figure 7-1) is assumed to be linear, price elasticities must necessarily differ at every point on the demand curve. However, the two price elasticities are assumed to be constant over the interval $Q_0 - Q_1$ in figure 7-1. It is further assumed that the outbreak-induced shift in demand leaves the new demand curve (D_1) parallel to the initial demand schedule (D), and that the supply curve during the outbreak shifts from S to S_1. It is also assumed that the initial price of restaurant services (P_0) equals the average check per customer—$3.52 for restaurants that sold only food and $10.97 for those that sold food and alcoholic beverages (National Restaurant Association, 1984). The initial quantity (Q_0) is the total number of customers that would have been served by the restaurants had the outbreak not occurred, estimated from survey data on normal weekly servings multiplied by the number of weeks of the outbreak. The actual number of customers served during the outbreak (Q_1) is estimated from Q_0 using the percentage drop in sales (from the questionnaire). The price on the demand curve, corresponding to the new quantity Q_1, is given by

$$p^* = P_0 + \frac{P_0\left(\dfrac{Q_1 - Q_0}{Q_0}\right)}{E} \qquad (7\text{-}1)$$

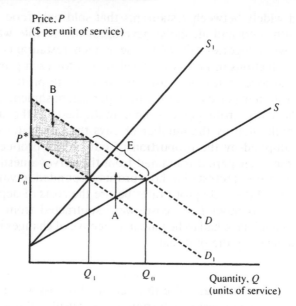

Figure 7-1. Loss of consumer surplus in the restaurant market due to the outbreak.

Note: Area A + B represents the loss in consumer surplus. The elasticity, E, is defined as

$$E = \frac{Q_1 - Q_0}{Q_0} \Big/ \frac{P^* - P_0}{P_0}$$

and is assumed to be either -2.2 or -7.8 over the interval $Q_0 - Q_1$.

where E is the elasticity of demand (assumed in the analysis to be -2.2 or -7.8). p^* is the equilibrium price if prices had been raised during the outbreak period. It is not the new price. Referring to figure 7-1, the loss in consumer surplus is represented by the triangle labeled A, plus the parallelogram labeled B. This loss may be calculated as follows:

Loss in consumer surplus =

$$\tfrac{1}{2}(P^* - P_0)(Q_0 - Q_1) + (P^* - P_0)(Q_1) \qquad (7\text{-}2)$$

Average Costs to Restaurants and Bars (Direct Costs and Lost Profits)

Table 7-1 provides the average costs to restaurants, bars, and restaurants and bars combined for three scenarios. This table covers only the first two components of losses in social welfare due to an outbreak—the direct costs of avoiding contaminated water and the lost profits. The third component, loss in consumer surplus, is discussed in a following section. Separate estimates of losses are provided for the averting activities, equipment purchases, and lost profits for each scenario in table 7-1.

Average losses due to averting activities of all restaurants and bars ranged from $2,227 to $4,006, or from 6.86 cents to 12.33 cents per customer. While restaurants faced higher costs than bars overall, their average costs were much lower on a per customer basis. Average lost profits for restaurants were $3,242. Information to make this calculation for bars was unavailable. Restaurants that sold only food averaged $2,547 in lost profits; the average for restaurants that sold food and drink was $3,585. (These losses are not reported in table 7-1.) Total losses to all restaurants amounted to 16.65 cents per customer under scenario A and 13.81 cents per customer under scenario B.

The low-skill wage rate of $5.00 per hour, assumed in scenario C, raised averting costs 18 percent above costs when the minimum wage rate is assumed (scenario A). The assumption that boiling water was combined with another activity, and, therefore, that time costs of boiling water were zero, lowers the average costs of averting activities by more than 34 percent. We believe that scenario B incorporates the most realistic assumptions about the averting behavior of restaurants and bars in the affected area.

Costs varied widely from establishment to establishment. Using the assumptions in scenario B, we estimate that the largest fast food restaurant in the sample (6,500 customers each week) spent an estimated 4.3 cents per customer to avoid contaminated water, while the most expensive restaurant reported costs of 40.1 cents per customer. The average cost per customer under scenario B is 6.06 cents.

Restaurants and Bars—Direct Costs and Lost Profits

Before the average costs for the sample of establishments can be applied to all eating and drinking establishments in the outbreak area to obtain an estimate of the total costs (total losses in producer surplus), the representativeness of the sample must be examined. Because a questionnaire was sent to *all* the restaurants and bars in the outbreak area, bias cannot be introduced into the estimate of losses through the sampling procedures. If a bias exists at all, it would have to be introduced through the response.

We cannot deduce even the direction of a response bias. One might expect managers of establishments most affected by the outbreak to have the greatest motivation to return the questionnaire. Conversely, one might also expect managers of establishments least affected by the outbreak to be more responsive because it would take them less time than more affected establishments to complete the questionnaire.

The size of an establishment might also influence the probability of its response, although in an uncertain direction. Larger establishments most likely keep better records and have available less expensive personnel to complete the questionnaire. Smaller establishments, on the other hand, might have less information to gather. Nevertheless, we could and did examine whether the size distribution of the respondents in terms of the number of employees was similar to the size distribution of all establishments in Luzerne County (*County Business Patterns 1980*, U.S. Bureau of the Census, 1982b). A chi-square test produced no statistically significant difference. Thus, the hypothesis of response bias based on the size of establishment was rejected.

Despite that result, the possibility that the estimates of average costs in table 7-1 were not representative, in some way, of all the affected restaurants and bars could not be ruled out. If there is a bias, however, its direction is not known.

The total costs of averting activities taken by the restaurants and bars in the affected area were determined by multiplying the average cost per establishment by the appropriate number of establishments affected, for each scenario. To determine total lost profits, average lost profits per customer were multiplied by the outbreak-induced reduction in the total number of customers served by restaurants in the affected area. This procedure was applied separately to restaurants serving food only, restaurants serving food and drink, and bars.

The next problem was to determine how many establishments fell into each of these categories. Data from the U.S. Bureau of the Census (1980) showed that 85 percent of the eating and drinking establishments in Luzerne County were primarily eating establishments and 15 percent were primarily bars. Assuming those percentages applied to the 250 establishments in the outbreak area, there were 213 restaurants and 37 bars. (These data may be compared with the proportion of establishments that returned the questionnaire—77 percent were restaurants and 23 percent were bars.) Based on a yearly survey of its membership, the National Restaurant Association estimated that 32 percent of the restaurants in the United States served only food in 1983, while 68 percent served both food and alcoholic beverages (National Restaurant Association, 1984). Assuming the same proportion for the 213 restaurants means that 145 served food and alcohol while 68 served food only.

The estimated total costs for restaurants and bars in the outbreak area are reported in table 7-2. Under scenario B—the scenario that incorporates the most realistic assumptions—total averting costs were

Table 7-2. Total Costs to Restaurants and Bars in the Giardiasis Outbreak Area (millions of 1984 dollars)

	Scenario[a]							
	A			B			C	
Loss category	Restaurants	Bars	Combined	Restaurants	Bars	Combined		Combined
Averting losses	0.723	0.089	0.882	0.540	0.048	0.588		1.002
Lost profits[b]	0.691	0.120	0.811	0.691	0.120	0.811		0.811
Total	1.414	0.209	1.693	1.231	0.168	1.399		1.813

Note: Calculations are based on the estimated 213 restaurants and 37 bars in the outbreak area and on Bureau of the Census weights. (U.S. Bureau of the Census, *County Business Patterns 1980, Pennsylvania* (Washington. D.C.. Government Printing Office. 1982).)

[a] Scenario A assumes a wage rate of $3.55 per hour and that boiling and hauling water are not combined with other activities. Scenario B assumes a wage rate of $3.55 per hour and that boiling water is combined with another activity. Scenario C assumes a wage rate of $5.00 per hour and that boiling and hauling water are not combined with other activities.

[b] Lost profits for bars are estimated by assuming that profit losses per establishment equal those of restaurants.

$588,000, and lost profits due to the reduction in demand for restaurant services amounted to $811,000, assuming that average losses in profits are identical for bars and restaurants. Total losses for scenario B came to $1.399 million.

Losses in Consumer Surplus

Losses in consumer surplus associated with reductions in the demand for services of restaurants and bars during the boil-water order are calculated for two elasticities of demand: -2.2 and -7.8. These calculations assume that the market could be separated into restaurants that sold only food, restaurants that sold food and drink, and bars.[2] For both scenarios, the average (initial) price of a meal (P_0) is assumed from National Restaurant Association data to be $3.52 for establishments that serve only food and $10.97 for establishments that serve food and drink (National Restaurant Association, 1984). No data on bars are available. To estimate consumer losses, the average check for bars is assumed to be $5.00.

In addition to the average price of a meal and the demand elasticities, estimates of the number of customer-visits (Q_0) that would have been served in the absence of the outbreak and the number served during the outbreak (Q_1) are needed. Based on responses to the questionnaire, in the absence of the outbreak, each food-only restaurant would have served an average of 73,696 customers; each food and beverage restaurant, 25,639 customers; and each bar, 4,992 customers. Scaling these figures to all the establishments in the outbreak area, 8.804 million customers would have been served in the absence of the outbreak (Q_0), but only 8.139 million customers were actually served (Q_1). Thus, 665,000 fewer customers were served because of the outbreak.

The estimated losses in consumer surplus, reported in table 7-3, amount to $2.26 million under scenario A and $662,000 for scenario B, depending on the demand elasticity assumed. Implicit in these calculations is the assumption that consumers did not increase their use of restaurants and bars over "normal" levels after the outbreak was over or switch their patronage to establishments outside the outbreak area. The total losses in consumer surplus are 25.6 cents and 7.5 cents on a per customer-visit basis for scenarios A and B, respectively. Viewed across types of

---

[2]One food and beverage establishment in the sample said that it had raised prices as a result of outbreak-related cost increases. This restaurant is treated as if it had its own market. Figure 4-3(c) in chapter 4 is assumed to apply. Given that 1 of 17 in the sample raised its prices, we assumed that 9 of the 145 food and beverage establishments in the outbreak area raised their prices.

Table 7-3. Losses in Consumer Surplus Associated with Restaurants and Bars in the Giardiasis Outbreak Area

Losses	Type of establishment				
	Food only	Food and beverage	Bars	Other[a]	Total
Millions of customer visits	5.012	3.487	0.185	0.120	8.804
Scenario A $(E = -2.2)$[b]					
Number of establishments[a]	68	136	37	9	250
Total losses (millions of 1984 dollars)	0.345	1.842	0.043	0.026	2.256
Loss per customer-visit (1984 dollars)[c]	0.069	0.528	0.23	0.22	0.256
Scenario B $(E = -7.8)$[b]					
Number of establishments[a]	68	136	37	9	250
Total losses (millions of 1984 dollars)	0.098	0.520	0.018	0.026	0.662
Loss per customer-visit (1984 dollars)[c]	0.019	0.149	0.097	0.22	0.075

[a]The 9 food and beverage restaurants assumed to have raised their prices are classified under "Other." With 145 food and beverage restaurants in total, subtracting these 9 leaves 136.
[b]E represents elasticity of demand.
[c]Total losses were divided by the number of customers that would have been served in the absence of the outbreak.

establishments, the more numerous and relatively harder hit food and beverage establishments account for most of these costs.

Total Losses to Restaurants and Bars

Table 7-4 summarizes the three components of social losses for five alternative scenarios drawn from scenarios A, B, and C in table 7-2, and scenarios A and B in table 7-3. The total losses double over the five scenarios, from a low of $2.061 million to a high of $4.069 million. On a per customer-visit basis, the losses range from 23.4 cents to 46.2 cents.

The baseline figure of 8.8 million customers that would have been served in the outbreak area had the outbreak not occurred is critical to the estimates in table 7-4. Two data sources on restaurants were used to check the reasonableness of the restaurants' share of this figure— 8.6 million. According to the National Restaurant Association (1984), the typical person living in the eastern United States purchased meals away from home 3.5 times per week in 1983. Assuming that all age groups were included in this estimate and that the net flow of restaurant traffic across the borders of the outbreak area was zero, and given that

Table 7-4. Summary of Losses to Restaurants and Bars
in the Giardiasis Outbreak Area
(millions of 1984 dollars)

| Loss category | Scenario[a] | | | | |
	A_1	A_2	B_1	B_2	C
Averting losses[b]	0.882	0.882	0.588	0.588	1.002
Lost profits[c]	0.811	0.811	0.811	0.811	0.811
Losses in consumer surplus[d]	2.256	0.662	2.256	0.662	2.256
Total losses	3.949	2.355	3.655	2.061	4.069
Loss per customer visit (dollars)[e]	0.449	0.267	0.415	0.234	0.462

[a]Scenario A assumes a wage rate of $3.55 per hour and that boiling
and hauling water are not combined with other ordinary activities. Elasticity of demand is −2.2 for A_1: −7.8 for A_2. Scenario B assumes a wage
rate of $3.55 per hour and that boiling water is combined with another
activity. Elasticity of demand is −2.2 for B_1: −7.8 for B_2. Scenario C
assumes a wage rate of $5.00 per hour and that boiling and hauling water
are not combined with other activities. Elasticity of demand is −2.2.
[b]Losses from table 7-2.
[c]Losses from table 7-2.
[d]Losses from table 7-3.
[e]The total costs were divided by the number of customer visits that
would have been served in the absence of the outbreak (8.804 million).

the average length of the outbreak for restaurants in the sample was
18.4 weeks, the 75,000 people in the affected area were assumed to
have purchased 4.83 million meals. Assuming the National Restaurant
Association figure was true for restaurants in Luzerne County means
that restaurant attendance might have been 44 percent lower than the
8.6 million customer-visits used in the calculations.

Restaurant traffic can also be estimated another way. If the average
number of customers per week per employee in the sample (167) is
multiplied by the number of employees in Luzerne County eating establishments—5,090 employees (U.S. Bureau of the Census, 1982b)—
customer traffic in Luzerne County is estimated to be 850,000 customers
per week. Given the RFF estimate of 213 restaurants in the outbreak
area and the U.S. Bureau of the Census (1982b) estimate of 343 eating
places in Luzerne County, we multiply Luzerne County customer traffic
(850,000 per week) by 0.62 (213/343) to produce an estimate of 527,000
customers per week. Multiplying this number by the weighted average
number of weeks in the outbreak (18.4) produces the estimate that 9.7
million customers would have been served during the outbreak. Thus,
the estimate of 8.6 million customer-visits is bracketed by two alternative
estimates of customer traffic during the period.

The number of customers patronizing bars, on the other hand, is
almost certainly not an overestimate. The sample of five bars included

four small neighborhood bars serving 200 or fewer customers per week and one serving 1,000 customers per week.

Limitations of the Analysis

Of all the estimates of losses made in this study, those for restaurants and bars in the outbreak area are likely to have the largest error. The most important source of error arises from the small sample size. With less than a 10 percent response rate from the heterogeneous group of establishments that constitute this sector, the average estimates of losses to the nonresponders cannot be extrapolated with much confidence, although estimates of the costs of averting behavior for the establishments that responded are probably as accurate as those made for individuals.

Lost profits for both restaurants and bars were calculated using average-profit-per-seat data from the National Restaurant Association. The same data were applied to bars. Without specific data for bars, the estimates may not be reliable. However, because the ratio of affected bars to affected restaurants is low, even large errors in estimating losses to bars will not greatly affect estimates of total losses. That issue aside, the use of average profit rates rather than marginal profit rates, which were unavailable, biases the losses upward.

An upward bias is also introduced by the assumption that reductions in demand caused by the outbreak were not offset by increases in demand experienced in the restaurant market outside the outbreak area or at a later time. While there is no information available to gauge the extent of the former effect, the latter effect appears to be small. Restaurant sales fell 10.5 percent, on average, during the outbreak and bounced back 7.8 percent afterward. Further, when pre-outbreak sales were compared with post-outbreak sales, eight of the seventeen restaurants in the sample showed no change in sales. Six did not regain all their lost sales after the outbreak ended, and three had larger sales after the outbreak than before it began. One of these raised prices during the outbreak, which, all else being equal, would tend to increase revenues.

Acting to create a downward bias in the estimates of losses is our inability to account for lost productivity. Although 27 percent of respondents indicated that such losses were present, none of them raised prices to compensate for it. Thus, the productivity loss left no trace on the market and, therefore, could not be measured directly. Moreover, the industry profit data from the National Restaurant Association survey were too gross in form to permit calculation of lost productivity.

Finally, the assumptions made to calculate losses in consumer surplus—linear demand curves extending to the price axis, and parallel shifts in demand in response to the outbreak—impart an unknown bias to the estimates of losses.

LOSSES TO OTHER BUSINESSES

Other businesses in Luzerne County were also affected by the outbreak. Estimates of losses are made in this section, and summarized in table 7-5, for hospitals, nursing homes, dentists, and day-care centers.

Hospitals

Nesbitt Memorial Hospital in Kingston was the only hospital in the area affected by the outbreak. This hospital was required to provide uncontaminated water, ice, and food to its 200 patients. The director of the hospital estimated that out-of-pocket expenses amounted to $6,820. In addition, he estimated the amount of time he diverted from other activities and the time spent by his staff to distribute water to patients. We valued the time the director spent at these activities at $1,837.

Table 7-5. Estimate of Losses to Other Businesses and Schools in the Giardiasis Outbreak Area
(1984 dollars)

| Activity | Cost category | | | |
	Direct costs	Time premium[a]	Possible costs[b]	Total costs
Other businesses				
Hospitals	8,657	0	6,278	14,935
Nursing homes	8,930	2,939	11,050	22,919
Dentists	10,359	0	6,045	16,404
Day-care centers	8,079	1,518	0	9,597
Others	N.A.	N.A.	N.A.	N.A.
Subtotal	36,025	4,457	23,373	63,855
Schools				
Public	16,865			16,865
Private	7,303	0	0	7,303
Subtotal	24,168			24,168
Total: Other businesses and Schools	60,193	4,457	23,373	88,023

Note: N.A. = data were not available.
[a]Time premium is the value of the time used to boil water, assuming that such time per customer served is equivalent to that of restaurants.
[b]Possible costs are those costs estimated from very little or uncertain data. They include estimates of the value of the time nurses spent distributing water to patients and the costs of a second pressurized water unit purchased by dentists.

Assuming that nurses devoted two minutes a day to attending to the patients' special needs for water and ice and assuming nurses earned $20,000 per year, the value of the time they spent on providing a potable source of water was $6,278 to the hospital's costs, for an estimated total cost of $14,935. Because water can be provided to patients while nurses perform other tasks, little or no time may have been lost from other tasks. Thus, these are listed as "possible costs" in table 7-5. If it is assumed that no nursing time was "lost" due to the boil-water order, the total cost to Nesbitt Memorial Hospital amounts to $8,657 ($6,820 plus $1,837 = $8,657).

Nursing Homes

The three nursing homes in the affected area had to provide uncontaminated food, ice, and water to their patients. During the period of the boil-water order, 352 people were patients at these three nursing homes. Detailed information taken from a personal interview with the manager of one of these institutions provides an estimate of the costs not only of protecting patients from exposure to contaminated water, but also of meetings devoted to developing strategies for avoiding contamination and of the additional time spent by nurses and other staff to provide a safe water supply.

Assuming that unit costs per patient were constant across nursing homes and ignoring food preparation costs, the three nursing homes spent in total an estimated $1,326 to protect patients from exposure to contaminated water and $1,333 on meetings. Food preparation costs were estimated from the average costs incurred by the restaurants in the affected area. These costs were either 6.06 cents or 8.90 cents per person served (table 7-1), depending on whether other work was ignored during the time it took to boil the water. Each nursing home patient was assumed to be a "customer" three times a day during the fourteen weeks each nursing home was required to comply with the boil-water order. The additional cost of food preparation was either $6,271 or $9,210, depending on the assumption made about boiling the water. Thus, the estimated cost to nursing homes was $8,930 ($1,326 + $1,333 + $6,271 = $8,930) plus the value of time (time premium) to boil water, $2,939 ($9,210 − $6,271 = $2,939).

Applying the same time factors and salary scale used for the hospital, the extra time required of nurses and other nursing home staff was valued at $11,050. However, because safe water might have been provided by nurses and staff in the course of performing their usual duties, additional costs for this service might not have been incurred. Thus, these are listed as "possible costs" in table 7-5. Summing all costs yielded an estimated total loss for the three nursing homes of $22,919.

Dentists

Dentists faced special problems in complying with the boil-water order because their work requires pressurized water. Estimating the additional costs to dentists involved determining the number of affected dentists and the average cost per dentist. With 343,079 people living in Luzerne County and 178 dentists,[3] a single dentist served an average of 1,927 people. If this ratio is applied to the 75,000 people in the outbreak area, thirty-nine dentists were affected by the outbreak. Telephone interviews with four dentists (10 percent of the estimated total) and the Pennsylvania Dental Association provided information on the fixed and variable costs of providing pressurized water. The fixed cost was at least $155 per unit (including installation, maintenance, and removal). Assuming one unit per dentist, we estimated that the additional fixed costs to the thirty-nine affected dentists was $6,045. Assuming two units per dentist, the estimate of the total cost doubled to $12,090. The actual fixed cost would be between these two estimates. The cost of the second pressure unit is listed as a "possible cost" in table 7-5. In addition to the fixed costs, dentists had to buy bottled water. Several dentists estimated these costs to be $5 per week, which implies a variable cost of $4,314 for all the dentists in the outbreak area. Thus, the giardiasis outbreak cost dentists in the affected area somewhere between $10,359 and $16,404.

Day-Care Centers

Day-care centers also incurred costs for water, juice, food, and "wet-wipes." Fourteen preschool child-care centers were found in the affected area; the average size was estimated to be sixty-nine children. The estimate of 966 children in organized day-care centers may be compared with the 3,964 preschoolers living in the area in 1980 (U.S. Bureau of the Census, 1982a).

The Apple Tree Nursery School in Kingston provided useful information for estimating costs. This center incurred additional costs for purchasing premade juices and spring water—21 cents per child per week and 5 cents per child per week, respectively. Applying these unit costs to all children in day-care centers throughout the outbreak period (22.1 weeks) yields a total additional cost to day-care centers of $5,557. (These unit costs are much larger than those incurred by elementary and secondary schools, as reported below.)

There are other costs as well. Based on interviews with a number of day-care centers, we estimated that 25 percent of the children used

---

[3]Personal communication with JoAnne Lockey, Pennsylvania Dental Association, Harrisburg, Pa., November 2, 1984.

prewatered, packaged towels for washing hands. The cost of these "wet-wipes" was estimated to be $855. In addition, 50 percent of the day-care centers prepared and served food during this period. Estimates of these costs per child per week were based on the additional costs experienced by restaurants and bars during the period of the boil-water order. The additional cost per customer served (6.06 cents when the time for boiling water was not included and 8.90 cents when it was) included the cost of soda. If that cost is subtracted, the cost per customer for the remaining water and ice amounts to 3.12 cents and 5.96 cents, respectively. Assuming each child in those day-care centers serving food is a "customer" once a day for every weekday of the boil-water order, additional costs total $1,667 or $3,185 depending on whether boiling the water involves labor time. Total costs for the day-care centers were thus either $8,079 or $9,597. (The difference between these estimates—$1,518—is listed as a "time premium" in the summary of losses to other businesses and schools in table 7-5).

LOSSES TO PUBLIC AND PRIVATE SCHOOLS

All schools in the outbreak area were required to provide uncontaminated water and food for their students. Data provided by the three public school districts in the outbreak area enabled us to estimate the extra costs incurred by all schools. Data on costs per week and number of weeks on the boil-water order obtained from the Wyoming School District, along with Bureau of the Census data on the school-age population in the affected area, provided information for estimating the costs to the numerous private schools in the outbreak area.

The accompanying estimates of costs were provided by the three public school districts:

Pittston Area School District	$10,000
Wyoming Area School District	5,515
Wyoming Valley West School District	1,350
Total, Public Schools	16,865

The costs incurred by private schools were estimated assuming the cost per student per week was the same as that of the Wyoming Area School District—8.7 cents per student per week. It was assumed that all children aged 5 through 18 attend school. By subtracting the 2,700 children attending public schools in the Wyoming District from the 3,647 school-aged children living in the district (Bureau of the Census, 1982), it was estimated that 947 children, or 26 percent of the Wyoming Area District school-age population, attended private school. Assuming that this percentage was true for all communities in the affected area, 3,794 children out of a total of 14,593 school-age children were estimated to be in private schools in the affected area.

Assuming that private school students were distributed evenly throughout the affected population and using data on the number of days each community in the area was on the boil-water advisory, the average weeks on the advisory (22.126) was applied to the 3,794 private school students. Using a unit cost of 8.7 cents per student per week, the total cost to the private schools was estimated to be $7,303. The total costs to public and private elementary and secondary schools in the outbreak area were thus estimated to be $24,168. They are summarized in table 7-5.

LOSSES TO GOVERNMENT AGENCIES

Government personnel at the federal, state, and local levels devoted time to issues and responsibilities raised by the outbreak. The Pennsylvania Department of Health, through its epidemiological studies, was the first agency to become involved. The Department of Environmental Resources took primary responsibility for issuing and enforcing the boil-water advisory and order, while the municipalities, the Luzerne County Emergency Management Agency, and the Luzerne County Housing Authority helped to provide the affected communities with safe drinking water. The Pennsylvania Consumer Advocate Office was involved in the suit against the water supply utility, while the Centers for Disease Control in Atlanta assisted with the epidemiological survey. In addition, other federal and state government agencies, including the state legislature, were involved. More detail on the specific roles played by the various government agencies involved in the outbreak is presented in chapter 5.

The value of the time government employees spent on tasks related to the outbreak, the costs of providing water to shut-ins, and travel expenses were added together to estimate the losses to government agencies due to the outbreak. (We assume that government employees would have had other tasks to do in the absence of the outbreak. See chapter 4 for a more complete discussion of the value of time spent by government employees on the outbreak.) For this estimate, it was necessary to determine the wage rate and the time each government employee devoted to the outbreak.

Most of this information was collected through a questionnaire sent to several key government officials (see appendix 7-B) and through follow-up telephone interviews. Some imputations and simplifying assumptions were made to fill in the gaps. For example, by assuming all municipalities incurred the same unit cost per capita served, detailed data collected from seven municipalities were used to estimate the costs incurred by the other eleven municipalities. While salaries of those who work for the state are a matter of public record, and thus were easily obtained, salaries for nonstate workers were not available. They were estimated from knowledge of the state government salary structure and the local wage rates.

Estimates of the labor hours spent on the outbreak and of the labor costs of government units—municipalities, counties, state agencies, and federal agencies—are presented in table 7-6. If a municipality or government agency provided information on nonlabor costs, it is included in the cost estimates. Most of the costs reported in the table are labor costs.

The total cost of the outbreak to the seven municipal governments in the survey was estimated to be $76,404. Based on the per capita costs in those municipalities and a population of 75,000 in the affected area, the total cost to all eighteen municipal governments was estimated to be $103,528. This procedure might have overestimated the costs to the eleven smaller municipalities that were not included in the survey. Because data to estimate these costs were not available, the estimate of $27,124 is listed in table 7-6 as a possible cost.

The estimate of total costs to *all* units of government—$254,648— may be an underestimate despite the small overestimate of the municipal costs. The value of time of many government employees, such as managers, secretaries, and clerks, who also played important roles in the outbreak, was not included in the estimate. Collectively, they may have devoted as many hours to the outbreak as did those with principal responsibilities, who were included in the estimates of losses. In addition, estimates were not made of the time spent by staff at the Environmental Protection Agency's Region III Office in Philadelphia, the EPA laboratory in Cincinnati, the U.S. Department of Agriculture, the Centers for Disease Control in Atlanta (with the exception of one epidemiologist), the Pennsylvania Public Utility Commission, the Pennsylvania State Legislature, and the Governor's Office in Harrisburg.

LOSSES TO THE WATER SUPPLY UTILITY

The cost of temporary measures taken by the Pennsylvania Gas and Water Company (PG&W) to provide the affected area with a safe water supply are social costs of the outbreak, as are the value of the time that staff spent at meetings, hearings, briefings, and strategy sessions, and other activities related to these temporary measures. Both should be included in the losses due to the outbreak.

Table 7-6. Estimates of Losses to Government Agencies Due to the Outbreak

Government agencies/units	Labor hours	Total costs (1984 dollars)
Municipalities in affected area:		
Municipalities surveyed		
Pittston City	1,610	17,050[a]
Forty Fort Borough	241	3,759
Wyoming Borough	525	4,200[b]
Moosic Borough	1,674	17,728[a]
West Pittston Borough	1,520	14,600
Kingston Borough	36	3,500[a,b]
Old Forge Borough	1,470	15,567[a]
Subtotal	7,076	76,404
Municipalities not surveyed (possible costs)		27,124[c]
Total costs: municipalities		103,528
Luzerne County Housing Authority	539	5,708
Luzerne County Emergency Management Agency	400	4,164[d]
State of Pennsylvania		
Department of Environmental Resources	6,412[e]	98,364
Department of Health	1,616	21,690[f]
Consumer Advocate Office	500	15,915[g]
Federal government		
Centers for Disease Control	190	5,279[h]
Others[i]	N.A.	N.A.
Total cost: government agencies		254,648

Note: N.A. = data were not available.

[a]Labor hours and labor costs from Forty Fort and Wyoming boroughs were used to derive an average labor cost of $10.59 per hour. This labor rate was used to compute labor costs for municipalities that provided only total labor hours.

[b]The May 23, 1984, issue of the *Sunday Independent* (Wilkes-Barre, Pennsylvania) reported costs of $4,200 for Wyoming Borough and $3,500 for Kingston Borough, figures that included the costs of water distribution to municipal offices. We assume that of the $3,500 for Kingston, $3,170 was spent to provide water to municipal offices, since Kingston reported expenses of only $330 for labor used to distribute water.

[c]Estimate is based on the per capita cost incurred by the seven municipalities in the survey and on the affected population in the remaining eleven communities.

[d]Only the director is included.

[e]Includes labor hours of the regional director, sanitarian, regional manager, community relations coordinator, and inspectors.

[f]Labor hours of the director, assistant director, three epidemiologists, and field nurses are included.

[g]Expenses for salaries, expert witnesses, travel, and lodging are included.

[h]One epidemiologist is included.

No data are available for hours devoted to the outbreak by staff of the Public Utility Commission, the Governor's Office, EPA Region III office, the EPA laboratory in Cincinnati, the U.S. Department of Agriculture, or state legislators.

PG&W incurred costs when it provided an alternative source of water supply for approximately 50,000 residents in the affected area by diverting water from an uncontaminated water supply to portions of the affected service area. The utility also constructed a 15,000-foot pipeline from Nesbitt Storage Reservoir around the contaminated Spring Brook Intake Reservoir. Finally, PG&W provided potable water at a number of distribution points throughout the area, retained consultants on the life cycle and transmission of *Giardia*, trapped and tested beavers for *Giardia* cysts, removed beavers from Spring Brook Intake Reservoir, cut down trees near the reservoir to discourage beavers from returning, and monitored water quality more closely.

The estimated cost of constructing the pipeline was $1.8 million (PG&W, 1984a). The cost of installing valves and caps to divert water from a clean reservoir to part of the affected area was $37,573 (PG&W, 1984b). Other costs incurred by PG&W such as consultant fees and laboratory tests were not available.[4]

REFERENCES

National Restaurant Association. 1980. *Food Service Trends* vol. 2, no. 4.
————. 1981. *Food Service Trends* vol. 3, no. 4.
————. 1984. *Restaurant Industry Operations Report '84* (Washington, D.C., National Restaurant Association and Laventhol and Horwath).
Pennsylvania Gas and Water Company. 1984a. Testimony presented before the Pennsylvania House Conservation Committee by J. Glenn Gooch, president and chief executive officer, Pennsylvania Gas and Water Company, Harrisburg, Pa., March.
————. 1984b. Testimony presented at the Pennsylvania Public Utility Commission Hearings, Exhibit 125, Item 2, by Joseph Lubinski, representing the Pennsylvania Gas and Water Company, Harrisburg, Pa., April–June.
U.S. Bureau of the Census. 1982a. *1980 Census of the Population: General Population Characteristics, Pennsylvania*, U.S. Department of Commerce (Washington, D.C., Government Printing Office).
————. 1982b. *County Business Patterns 1980, Pennsylvania*, U.S. Department of Commerce (Washington, D.C., U.S. Government Printing Office).

---

[4]We have requested cost estimates from PG&W, but as of this writing, they have not been provided.

Appendix 7-A

Questionnaire for Restaurants and Bars

Note: This questionnaire was designed by Resources for the Future for use in the survey of 250 restaurants and bars in the giardiasis outbreak area, Luzerne County, Pennsylvania, in 1984.

Response of Eating and Drinking Establishments to the Giardiasis Outbreak, Luzerne County, 1983–84

Establishment name: _____

I. Type and size of establishment:

1. Is your establishment primarily a restaurant _____ bar _____ ?

2. Please indicate the number of employees:
   1–4 _____   20–49 _____
   5–9 _____   50–99 _____
   10–19 _____   100 or more _____

3. Please indicate the average number of customers you serve per week: _____ customers per week.

II. Location of establishment and boil-water advisory:

4. Is your establishment on the Spring Brook Reservoir water supply system?   Yes _____   No _____

   If you answered "No," do not complete the rest of this questionnaire.

5. The boil-water advisory was issued on December 23, 1983. When was the boil-water advisory lifted for your establishment? _____ Did you continue to take any specific precautions after the boil-water advisory was lifted?   Yes _____   No _____   Until what date? _____

III. Alternative water supplies:

6. Which of the following sources of water did you use during the outbreak? Check as appropriate.

   a. Purchased bottled water   _____
   b. Boiled water   _____
   c. Obtained water from distribution
      points established by local
      governments and PG&W   _____
   d. Obtained water from a source
      outside the affected area   _____
   e. Other (specify)   _____   _____

   For each alternative checked, please answer the corresponding set of questions below:

   a. Bottled water

7. On average, how much additional bottled water did you buy each week? _____ gallons. For how many weeks? _____ weeks.

8. What price were you charged? $_____ per gallon

9. Was the water delivered to you? Yes _____   No _____

10. If not, how many trips per week did your employees make to get bottled water? _____ trips

11. What was the time required for each trip? _____ minutes

    b. Boiled water

12. On average, how much water did you boil per week? _____ gallons. For how many weeks? _____ weeks.

13. What kind of fuel was used? _____ gas heat   _____ electric heat

14. Did this activity take additional time of your employees or were they able to do other work while boiling water? _____

15. Did you have to purchase containers for this purpose?
    Yes _____   No _____

    c. or d. Water from distribution points or other sources outside the affected area

16. On average, how much uncontaminated water per week did you obtain from distribution points or from sources outside the affected area? _____ gallons

17. How many trips per week did your employees make for this purpose?
_____ trips. For how many weeks? _____ weeks

18. What was the time required for each trip? _____ minutes
Did you have to purchase containers for this purpose? Yes _____
No _____

IV. Ice:

19. Did you purchase ice made with an uncontaminated source of water during the outbreak? Yes _____ No _____

20. If so, on average, how much ice per week did you buy? _____ pounds

21. What price did you pay for ice? $_____ per pound

V. Fountain beverages:

22. Did you substitute bottled or canned soft drinks, soda, or mixers for fountain drinks? Yes _____ No _____ For how many weeks? _____ weeks

23. If so, how many additional cases of bottled or canned beverages did you buy each week for this purpose? _____ cases. What size were the bottles or cans (e.g., 12-oz. bottles)? _____ oz.

24. What was the cost per case? $_____ per case

VI. Other operating costs and investment purchases:

25. Did the outbreak cause you to lay off or hire additional workers?
Yes _____ (Laid off _____ , Hired _____); No _____

26. If so, how many? _____ employees.
Over what period? _____ weeks.

27. Did the outbreak cause you to purchase capital equipment, such as water purifiers or high temperature dishwashers?
Yes _____ No _____

28. If so, please list these purchases and their approximate cost:
_____
_____
_____

VII. Other effects:

29. Did the outbreak have any effect on your sales?
    Yes _____   No _____

30. If so, what would you estimate the change in sales to be, compared with "normal" sales (say, the sales volume experienced one year earlier)?

    Increased more than 19%   _____
    Increased 10–19%   _____
    Increased 1–9%   _____

    Decreased 1–9%   _____
    Decreased 10–19%   _____
    Decreased more than 19%   _____

31. Did you change food or drink prices as a result of the outbreak?
    Yes _____   No _____

32. If so, how much?

	Food prices	Drink prices
Increased more than 19%	_____	_____
Increased 10–19%	_____	_____
Increased 1–9%	_____	_____
Decreased 1–9%	_____	_____
Decreased 10–19%	_____	_____
Decreased more than 19%	_____	_____

33. After the boil-water advisory was lifted, did you notice a change in sales?   Yes _____   No _____

34. If so, what would you estimate the change in sales to be, compared with sales during the outbreak?

    Increased more than 19%   _____
    Increased 10–19%   _____
    Increased 1–9%   _____

    Decreased 1–9%   _____
    Decreased 10–19%   _____
    Decreased more than 19%   _____

35. Did the outbreak have any other effects on your business that might be relevant to an economic assessment of the impacts of contaminated public water supply? If so, please describe them below:

    _____
    _____
    _____

Appendix 7-B

Questionnaire for Government Agencies

Note: This questionnaire was designed by Resources for the Future for the survey of government officials involved with the giardiasis outbreak in Luzerne County, Pennsylvania, in 1984.

Questionnaire
Kingston–Pittston Outbreak

Name _____

Agency _____

Position _____

I. <u>Activities</u>

List the principal activities you engaged in with respect to the outbreak (e.g., meetings attended and epidemiological surveys conducted, etc.).

_____

_____

_____

_____

_____

II. <u>Time Cost</u>

We would like your estimate of the work time you devoted to the Kingston-Pittston outbreak. Also, please note the date you began to work on this issue and the date you stopped (write "continuing," if applicable). To jog your memory, key dates are:

<u>December 23, 1983</u>:   The boil-water advisory issued (for the Kingston-Pittston giardiasis outbreak);

January 11, 1984:	Public meeting held with DER, Department of Public Health, PG&W, and others;
February 13:	DER conveys orders to PG&W;
March 30 and April 9:	Most areas are removed from the advisory;
March 28–29:	House hearings;
April 24–26:	Senate hearings.

Month	Average number of hours worked per week (include overtime)	Percentage of this time devoted to the outbreak
November 1983	_____	_____
December 1983	_____	_____
January 1984	_____	_____
February 1984	_____	_____
March 1984	_____	_____
April 1984	_____	_____
May 1984	_____	_____
June 1984	_____	_____
July 1984	_____	_____

How many hours do you normally work per week? _____

III. Associated Expenses

A. If your primary place of work was outside the outbreak area, please note the number of trips you took to the area and estimate the average cost of food and lodging per trip:

Number of round trips _____.

Average cost of food and lodging per trip _____.

B. Transportation:

If public transportation, average cost per round trip: _____

If you drove to and from the outbreak area, how many people were in the car with you? _____ .

IV. Please list the names and positions of people outside your agency with whom you had frequent contacts concerning your work on the Kingston-Pittston outbreak.

_____

_____

_____

_____

8 / Summary of Losses and Policy Implications for Luzerne County

This final chapter on the case study summarizes the losses resulting from the outbreak and discusses the implications of these losses for protecting the residents and businesses in the affected area from the risk of future outbreaks. The first section summarizes the estimates of losses to individuals and to businesses and government agencies, as reported in chapters 6 and 7, and sets out the limitations of the analysis, indicating, where possible, whether those limitations result in overestimates or underestimates of losses. In the second section, the benefits of avoiding future outbreaks of giardiasis in the affected area are compared with the costs of providing filtration of the raw water supplied to that area from the Nesbitt and Spring Brook Intake Reservoirs. The chapter concludes with a statement concerning the use of benefit analysis and benefit–cost analysis in decisions to protect the public from drinking water contaminated with pathogenic organisms.

TOTAL LOSSES DUE TO THE OUTBREAK

Estimates of the losses resulting from the 1983 Luzerne County outbreak are summarized in table 8-1. These losses are divided among four major categories: (1) losses due to illness; (2) those due to actions taken by individuals and businesses to avoid contaminated water; (3) those due

Table 8-1.   Summary of the Losses Attributable to the Outbreak of Giardiasis in Luzerne County, Pennsylvania
(millions of 1984 dollars)

Loss category	(High estimate) A	B	(Low estimate) C
	Composite scenarios		
GROUP I			
Losses due to illness[a]			
Medical costs	1.07	1.05	1.03
Loss of work	2.15	1.63	1.25
Losses due to averting actions			
Individuals—lower bound[a]	19.41	4.57	4.24
Restaurants and bars[b]	1.00	0.59	0.59
Schools and other businesses[c]			
(direct costs)	0.06	0.06	0.06
Losses to government agencies[d]	0.23	0.23	0.23
Losses to water supply utility	1.84	1.84	1.84
Subtotal, Group I	25.76	9.97	9.24
GROUP II			
Losses due to illness			
Loss of productivity[a]	2.22	1.89	1.67
Loss of leisure time[a]	5.26	3.91	2.95
Losses due to averting actions			
Individuals			
Best estimate minus lower bound[a]	19.10	8.37	7.88
Restaurants and bars			
Lost profits[b]	0.81	0.81	0.81
Consumer surplus losses[b]	2.26	2.26	0.66
Schools and other businesses			
Time premium and possible costs[c]	0.03	0.03	0.03

(continued)

Note: Estimates of losses shown in this table are presented for two confidence levels—group I and group II. These levels are used to organize losses on the basis of the confidence in the underlying assumptions, data, and methods of analysis used to estimate them. The actual losses are at least as great as the totals reported for the group I losses, resulting in a relatively high degree of confidence in these estimates as a lower bound. Although the losses in group II are real, there is less confidence in these estimates.

[a]From table 6–7.
[b]From table 7–4.
[c]From table 7–5.
[d]From table 7–6.

Table 8-1 (continued)

Loss category	Composite scenarios		
	(High estimate) A	B	(Low estimate) C
GROUP II (*continued*)			
Losses to government agencies			
Possible costs[d]	0.03	0.03	0.03
Subtotal, Group II	29.71	17.30	14.03
TOTAL, GROUPS I and II	55.47	27.27	23.27
GROUP III			
Losses due to:			
All other tangibles[e]	Not estimated		
Intangibles[f]	Not estimated		

[e]All other tangibles not included in groups I or II. They include highly valued leisure activities, costs of legal fees, costs of misdiagnosis of giardiasis, losses to businesses due to reductions in productivity (in addition to those reflected by individual losses in productivity included in group II), net losses to individuals resulting from substituting more expensive beverages for those that require tap water, the value of time devoted to the outbreak by some government personnel (in addition to that already included in groups I and II), and the effects on businesses in the outbreak area that were not investigated (e.g., hotels, motels, and meat-packers).

[f]Intangibles include pain and suffering of those who were ill, anxiety of those living in the outbreak area over the possibility of contracting the disease, and the diminished intrinsic value resulting from the loss of a pure water supply for drinking, food preparation, and personal hygiene.

to the time spent by government officials on the outbreak, and related expenses; and (4) those incurred by the water supply utility. In addition, several categories of losses, both tangible and intangible, are included in table 8-1 to remind the reader that this study did not attempt to make estimates for all possible categories of loss.

Estimates of losses in table 8-1 are presented for three composite scenarios, which are described in table 8-2. A composite scenario represents an integration of the scenarios presented in chapters 6 and 7 for major loss categories, based on the assumptions used to calculate losses. Thus, composite scenario A incorporates, in part, scenario A losses due to illness (table 6-7) and scenario A losses due to individual averting behavior (table 6-7) because the assumptions used in developing those losses are the same. In composite scenario A, the value of time for nonworking adults is assumed to be $6.39 per hour, the average after-tax wage of those in the sample of confirmed cases working outside the home. (All wage rates and prices are reported in 1984 dollars.) In composite scenario B, the value of time for nonworking adults is taken to be $2.65 per hour, the minimum wage after an allowance has been taken

Table 8-2. Descriptions of Composite Scenarios in Table 8-1

Loss category	Composite scenarios (High estimate) A	B	(Low estimate) C
Losses due to illness (scenario/table)	A/6-7	B/6-7	C/6-7
Hourly wage[a]			
All workers	Survey[b]	Survey[b]	Survey[b]
Homemakers, retirees, unemployed	$6.39	$2.65	0
Others	0	0	0
Losses due to averting actions (scenario/table)	A/6-7	B/6-7	C/6-7
Individuals			
Hourly wage			
Adult workers	$6.39	$6.39	$6.39
Homemakers, retirees, unemployed	$6.39	$2.65	0
Teenage workers	$2.65	$2.65	$2.65
Others	0	0	0
Price per gallon, bottled water[c]	$1.17	$0.66	$0.66
Boiling water[d]	Sole activity	Combined	Combined
Hauling water[d]	Sole activity	Sole activity	Sole activity
Restaurants and bars (scenario/table)	C/7-4	B_1/7-4	B_2/7-4
Price elasticity of demand[e]	−2.2	−2.2	−7.8
Hourly wage	$5.00	$3.35	$3.35
Boiling water[d]	Sole activity	Combined	Combined
Hauling water[d]	Sole activity	Sole activity	Sole activity
Schools and other businesses (table)	7-5	7-5	7-5
Loss estimates	Survey[b]	Survey[b]	Survey[b]
Losses to government agencies (table)	7-6	7-6	7-6
Loss estimates	Survey[b]	Survey[b]	Survey[b]
Losses to water supply utility (chapter 7, text)			
Loss estimates	Survey[b]	Survey[b]	Survey[b]

[a]Hourly wages for workers during working hours are before-tax. Hourly wages for workers during leisure time and all others are after-tax (the before-tax wage rate times 0.79). The value of leisure time is the after-tax wage rate for both workers and all others. Wage rates are in 1984 dollars.

[b]"Survey" means that the cost item is taken or calculated from one of the RFF surveys (see chapters 6 and 7) of confirmed cases of giardiasis, households in the affected area, restaurants and bars in the affected area, and government agencies involved in the outbreak.

[c]The delivered price of a gallon of bottled water is $1.17. The purchase price is $0.66. Thus, in composite scenario A, the time cost for purchasing bottled water at a store is assumed to be $0.51 per gallon ($1.17 − $0.66 = 0.51). Prices are in 1984 dollars.

[d]Assumptions about combined activities affect the calculations of time loss for boiling and hauling water. An averting activity combined with a regular activity is assumed to have no time cost. If the activity of boiling or hauling water is performed along with a nonaverting activity, in the sense that the performance (e.g., time to complete, the enjoyment, the quality of product) of the nonaverting activity is unaffected, the boiling or hauling activity is considered to be a combined activity. Otherwise, it is a sole activity. Surveyors accepted whatever a household reported about hauling water. Sometimes, however, when a household reported that boiling water was not combined with another activity, we assumed that it was (composite scenarios B and C). See text in chapter 6 for a full explanation.

[e]The price elasticity of demand is the percentage change in quantity demanded for a 1 percent change in price. See chapter 7 for a detailed explanation.

out for taxes. In composite scenario C, the value of time for home-makers, retirees, and unemployed adults is assumed to be zero. For workers in all three composite scenarios, the value of time is taken to be the wage rate. For time losses associated with illness, the wage used is the actual reported wage from the confirmed case survey, taken before or after taxes, depending on whether the time lost to illness is assumed to be work time or leisure time. For averting losses, the value of time for all workers is the average after-tax wage from the confirmed-case survey.

The scenarios used to estimate losses associated with restaurants and bars are not as easily integrated into this scheme. Table 7-4 presents five alternative scenarios for restaurants and bars. To ensure a manageable presentation, three of them were used: the highest-cost estimate (C), the lowest-cost estimate (B_2), and one in between (B_1). Scenario B_1 has assumptions that match as closely as possible those of scenario B for losses due to illness and losses due to individual averting behavior. Scenario C matches the high-loss A scenario for individual losses, and scenario B_2 matches the low-loss C scenario for individual losses. Because there is only one scenario for losses to schools, other businesses, government agencies, and the water supply utility, no further complications arise in constructing the three composite scenarios.

Estimates of losses in table 8-1 are also presented for two "confidence" levels—group I and group II. These levels are used to organize losses on the basis of the confidence we have in the underlying assumptions, data, and methods of analysis used to estimate them. These two levels also are used in the benefit–cost analysis presented in the second part of this chapter, to organize the benefits of constructing a filtration plant near Spring Brook Intake Reservoir on the basis of the kinds of benefits involved. This organization of benefits reflects more the practical problems involved in using benefit–cost analysis to inform public policy on matters of public health than it does the measurement problems, although the measurement problems exacerbate the practical problems. This requires a brief explanation.

Financial (tangible) losses are more easily understood by society than nonfinancial (intangible) losses. Because of this, there will be little disagreement that the actual losses associated with an outbreak are at least as great as the financial losses. This stems in part from legal tradition in assessing damages and in part from the more direct effects of financial losses experienced by individuals, businesses, and government. If a filtration plant can be justified on the basis of those categories of benefits where there is little disagreement on the estimates, there should be little disagreement on the guidance for policy provided by the benefit–cost analysis. If, on the other hand, the filtration plant cannot be justified on the basis of the more tangible benefits in group I, it will be necessary to incorporate in the totals some, or all, of the less tangible categories of benefits in group II. This promises to be more controversial, first because of the nature of the loss (benefit) categories in group II, and second because there is less confidence in the estimates of the losses in this group.

The actual losses are at least as great as the totals reported for the group I losses, resulting in a relatively high degree of confidence in these estimates as a lower bound. We have less confidence in the group II estimates of losses. For example, the direct expenditures that restaurants and bars made to avoid contaminated water are included in the group I estimates because the essential data used in these calculations were obtained directly from some of the affected establishments. However, losses in consumer surplus in the restaurant and bar market are included in the group II estimates because of uncertainty about the shapes of the demand functions for the services of restaurants and bars in Luzerne County.

The losses due to illness for both the group I and II estimates omit hospital costs for the clinical cases of giardiasis (estimated to be 5,630 cases) on the assumption that only confirmed cases of giardiasis were treated in hospitals. Losses in productivity and in leisure time activities are included in the group II estimates. The estimates of losses for schools and other businesses in group I exclude some rather speculative estimates for time losses, which are, however, included in the group II estimates. Also, some speculative estimates of costs to municipal governments based on unit cost estimates from other municipalities appear only in the group II estimates of losses.

The lower-bound estimate of losses due to actions taken by individuals to avoid contaminated water is included in the group I estimates of losses. This is a firm lower bound; actual losses are not likely to be less than this estimate. It is based on the actual costs to households of providing uncontaminated water for drinking, food preparation, and personal hygiene.

The difference between the best estimate of losses due to actions taken by individuals to avoid contaminated water and the lower-bound estimate (table 6-7) is included in the group II estimates of losses. This additional loss over the lower-bound estimate is the remainder of the consumer surplus lost from an increase in the price of uncontaminated drinking water (figure 6-1). Because there was no information on the shape of the demand function for water for drinking, food preparation, and personal hygiene, it was assumed to be linear between the quantity of uncontaminated (bottled, boiled, and/or hauled) water actually used and the quantity of tap water used for these purposes before the outbreak.

The estimates of the total losses—all of which are in 1984 dollars— of the Luzerne County outbreak range from a low of $9.2 million (scenario C, group I estimate) to a high of $55.5 million (scenario A, sum of group I and group II estimates), depending on the composite scenario selected and the confidence placed in the estimates of losses in group II. Differences in the estimates of the losses due to individual averting behavior account for the major portion of this range. Total (group I and II) individual averting behavior losses for scenario A are $38.5 million, for example, while those for scenario C in group I are only $4.2 million. The difference, $34.3 million, accounts for 74 percent of the range between the lowest and highest estimate of total losses of the outbreak.

Because losses due to averting behavior affect the estimates of total losses so significantly, they are analyzed in greater detail. Specifically, total individual averting behavior losses in group I and group II are disaggregated by whether the losses are out-of-pocket (for bottled water, gasoline, and cooking fuel) or implicit, in terms of the value of time involved. Such losses as a percentage of total losses due to individual averting behavior are presented in table 8-3.

Time losses dominate averting behavior losses in composite scenario A in table 8-1, constituting 82 percent of the group I losses and 63 percent of group II losses. Most of these losses are due to the time costs involved in boiling water (99 percent of the costs of boiling water), as reported by the households in the survey; yet, only one-third of the households in the sample reported that they incurred a time cost for this activity. Had a scenario been constructed which assumed that boiling water resulted in time costs for every household in the affected area, the absolute size of the losses due to averting behavior and the percentage of such losses caused by lost time would have been much larger.

Time losses of averting actions under scenarios B and C in table 8-3 are less significant because the value of time spent boiling water and buying bottled water is assumed to be zero.

For all three composite scenarios in table 8-1, the losses due to individual averting actions vary significantly depending on whether the lower-bound (group I) procedure or the best estimate (group II) procedure is used to value the losses. This is illustrated in table 8-4. The out-of-pocket losses based on the best-estimate procedure are at least 1.9 times those based on the lower-bound procedure, while time losses are from 0.75 to 1.60 times the corresponding lower-bound figure, depending on the scenario.

Limitations of Analyses

The loss estimates reported in table 8-1 should be interpreted in light of the caveats discussed in chapters 6 and 7. For convenience, these caveats are recapitulated here.

Table 8-3. Percentage of Out-of-Pocket Losses and Time Losses in Total Losses Due to Individual Averting Behavior
(percent)

Composite scenario	Group I losses				Group II losses	Group I plus group II losses
	Boil	Haul	Bottled	Total		
Scenario A						
Out-of-pocket losses	0.6	54.6	60.4	17.6	37.0	27.2
Time losses	99.4	45.4	39.6	82.4	63.0	72.8
Total losses	100.0	100.0	100.0	100.0	100.0	100.0
Scenario B						
Out-of-pocket losses	100.0	63.0	100.0	75.0	77.9	76.9
Time losses	0	37.0	0	25.0	22.1	23.1
Total losses	100.0	100.0	100.0	100.0	100.0	100.0
Scenario C						
Out-of-pocket losses	100.0	70.5	100.0	80.9	83.2	82.5
Time losses	0	29.5	0	19.1	16.8	17.5
Total losses	100.0	100.0	100.0	100.0	100.0	100.0

Note: Table is based on losses due to averting actions and on the three scenarios reported in tables 6-7 and 8-1 in this volume.

Table 8-4. Ratio of Group II Losses to Group I Losses for the Losses Due to Individual Averting Behavior

Type of loss	Composite scenario		
	A	B	C
Out-of-pocket losses	2.07	1.90	1.91
Time losses	0.75	1.62	1.63
Total averting losses	0.98	1.83	1.86

Note: Table is based on the lower-bound estimate and the best estimate of the losses due to individual averting actions, for the three scenarios presented in table 6–6 in this volume.

The principal limitation is the omission of some loss categories. These omissions bias total losses downward. Some loss categories were left unvalued because the losses were intangible; that is, they left no trace in data available on market and social behavior. These loss categories include pain, suffering, aggravation, anxiety, and diminished intrinsic value resulting from the loss of a "pure" water supply for drinking, food preparation, and personal hygiene.

The bias that results from omitting intangibles is likely to be large. In legal proceedings, lawyers typically multiply out-of-pocket losses by a factor of 2 or 3 to determine the value of pain and suffering.[1] While this legal rule of thumb may not be an appropriate measure for the willingness to pay to avoid pain and suffering, it does indicate the magnitude of the bias that may result from omitting intangibles.

For the Luzerne County giardiasis outbreak, losses due to pain and suffering may be low compared with losses due to aggravation and anxiety. Only the 6,000 people who contracted the disease experienced pain and suffering, whereas the entire population at risk—75,000 people—experienced some degree of aggravation and anxiety. Random telephone interviews revealed that people were anxious about the quality of the drinking water and remained so even after the outbreak ended. More than half of the fifty households in the RFF telephone sample continued to take some averting actions after the outbreak was over.

Had time and resources permitted, seven other unvalued loss categories could have been valued using objective measures of behavior. These were losses associated with highly valued leisure time activities; the costs of legal fees; the costs of misdiagnosis; productivity losses to businesses (in addition to those reflected by individual losses in productivity estimated in chapter 6); net losses to individuals associated with substituting less preferred (or more expensive) beverages for those that required tap water; the value of time devoted to the outbreak by some government personnel; and the effects on hotels, motels, meatpackers, and other businesses in the outbreak area that were not investigated.

The estimates of losses that are included involve various degrees of uncertainty, either because of the assumptions made or because of data limitations not serious enough to preclude an attempt at valuation. An examination of the most important assumptions and limitations finds that they generally lead to overestimates (though, because some losses are not estimated at all, total losses are probably underestimated).

The estimates of losses due to illness are based on two key assumptions: first, that the losses of an average clinical case of giardiasis are identical to those of an average confirmed case once hospital costs are subtracted; and, second, that the attack rate used is a reasonable measure of disease incidence in the affected population. Consider the first assumption. It is reasonable to expect that the average confirmed case was more severe than the average clinical case because those who were seriously ill were more likely to have sought confirmation of their illness. To the extent that the severity of illness differs, the estimates of the losses due to illness in table 8-1 may be overstated.

Turning to the second assumption, it is standard practice to describe an outbreak in terms of an attack rate derived from a random survey

---

[1]Personal communication from Scott J. Rubin, Pennsylvania Office of Consumer Advocate, Harrisburg, Pa., June 1984.

of the exposed population and a confirmed case rate. Used by itself, the attack rate derived from the random survey may overestimate the number of cases because it may include diseases that have giardiasis-like symptoms but are not giardiasis. Moreover, because it is derived from a sample, this attack rate may be unreliable if the sample size is not random or if it is too small.

Used by itself, the number of confirmed cases is almost certainly an underestimate of the actual number of cases, because patients may not follow through on all the steps required to confirm a clinical case. Confirmation of giardiasis requires that: (1) a person recognize that he or she may have a disease requiring either the care of a physician or the testing of a stool sample; (2) the disease not clear up before the patient seeks such care or submits a stool sample; (3) if consulted, a physician require the patient to provide a stool sample for examination; (4) the stool sample be properly prepared; (5) the laboratory find *Giardia* cysts in the stool; and (6) the health department be notified of the confirmed case.

We chose to use the attack rate of 8 percent, estimated by the Pennsylvania Department of Health, to scale the losses due to illness rather than the confirmed case rate of 0.5 percent. We decided that the former was the better indicator of the number of people who contracted giardiasis because, in principle, it includes both the clinical and confirmed cases. (For a more detailed discussion of the attack rate, see, in chapter 6, the section on "Total Losses to Individuals in the Outbreak Area.")

There were several reasons for this decision. First, the number of cases in which another disease masqueraded as giardiasis is likely to be low because an unusually stringent clinical definition of giardiasis was used—diarrhea lasting ten days or more. An RFF survey nine months later, which used the same clinical definition, yielded about the same attack rate. The sizes of both samples (233 in the Department of Health survey, 148 in the RFF survey) were large enough to hold sampling errors to reasonably low levels, and both samples were random.

Second, it is likely that many cases of giardiasis were not confirmed. The expense of a visit to a physician and subsequent stool analysis discourage these activities; moreover, a single stool analysis is not a foolproof test. Even with well-prepared samples from someone with an active case of giardiasis, up to three stool samples may be required to detect the cysts. Occasionally, symptoms are relieved by over-the-counter drugs, and those with the disease do not seek further medical treatment or analysis. Symptoms may also disappear without medication.

Perhaps the most important reasons for cases going unconfirmed, according to our research in other outbreak areas, is that physicians commonly provide patients with prescriptions based on clinical symptoms and do not seek a confirmed diagnosis; this is most likely to happen when one member of a family has a confirmed case of giardiasis and the other family members have the same symptoms. Although we do not know for sure, we suspect that physicians in Luzerne County did not try to confirm all the cases of giardiasis they treated. Finally, because Pennsylvania law requires physicians to report only confirmed cases, clinical cases are generally not reported to the Department of Health.

Limitations also apply to the estimates made for other loss categories. The group I estimates of the losses due to individual averting behavior may be high because the survey limited the mix of strategies that could be reported by respondents. Since more complex strategies would be used only if they reduced losses to individuals, limiting the mix of strategies in the survey may have imparted an upward bias to estimates of losses.

The assumption that reductions in the demand for the services of restaurants and bars in the affected area were not offset by increases in the demand for such services outside the affected area and after the outbreak ended imparts an upward bias to the estimates of losses for restaurants and bars. Data limitations for restaurants and bars in the affected area also could result in an overestimation of losses. Use of average profit rates, rather than the preferred, but unavailable, marginal profit rates, suggests that these estimates may be high too. The small sample of establishments (22 out of 250) and the assumptions of linear demand functions and parallel shifts in demand made to compute losses in consumer surplus have an unknown effect on estimates of losses.

Confidence in Estimates

Given the loss categories omitted, the caveats and limitations associated with those estimates that are included, and the biases these omissions and caveats impart, the reader may wonder how to interpret the estimates of losses presented in table 8-1.

We have a high degree of confidence in the estimates of losses to government agencies and the water supply utility reported in group I. These are firm lower bounds. The actual losses are likely to be at least as great and are almost certainly higher. Losses due to illness and to the averting behavior of individuals and businesses are not likely to be lower than those reported for scenario C in group I. In this scenario, work time losses of all but those who receive explicit wages are valued at zero, leisure time losses are valued at zero, and it is assumed that no one devotes special time to boiling water or makes a special trip to purchase bottled water. (However, special trips are made to obtain water from water distribution centers.) In our judgment, the group I estimate of losses for scenario C—$9.24 million—represents a firm lower bound on the total losses due to the outbreak in Luzerne County. Total losses are unlikely to be less than this amount and more likely to be substantially greater.

We have less confidence in the data and methods of analysis used to estimate the losses in group II. Nonetheless, the loss categories—losses in productivity, profits, leisure time activities, and consumer surplus— are unquestionably present and, therefore, raise losses above those reported in group I. Whether these group II losses are as large as those shown in table 8-1 cannot be resolved here. Since the differences between the best estimates and lower-bound estimates of losses due to individual averting behavior (table 6-7) are responsible for the major portion of the group II losses (ranging between 48 and 64 percent of these losses), judgments about where the actual losses for this group lie will largely determine the best estimate of the total losses of the outbreak.

One final point. This research focused exclusively on estimating the total losses from an outbreak of giardiasis and did not investigate the distribution of losses. The results provide some indication of who pays, in terms of the initial incidence on individuals, businesses, schools, government agencies, and the water supply utility. However, the initial incidence of such losses within each cost category was not examined by income level, for example, or age. Nor was the final incidence of these losses examined. While a distributional analysis of costs would be of interest to some, it would not serve the principal interests of valuing total social losses to a community from an outbreak of giardiasis.

POLICY IMPLICATIONS FOR LUZERNE COUNTY

The estimates of total losses due to the outbreak range between $9.2 million and $55.5 million (table 8-1). The issue for public policy is what more should be done, if anything, to protect the residents and businesses in the Spring Brook–Hillside Service Area from possible future outbreaks.

By fall 1984, the Pennsylvania Gas and Water Company (PG&W) had constructed a 15,000-foot pipeline from Nesbitt Reservoir around the contaminated Spring Brook Intake Reservoir to serve the area affected by the outbreak. The utility also had removed trees from the shoreline of the reservoir to discourage beavers from inhabiting the area, and it planned to drain and dredge Spring Brook Intake Reservoir to remove any *Giardia* cysts that lay in the bottom sediments. And the Pennsylvania Department of Environmental Resources had ordered PG&W to construct a filtration plant to treat raw water supplied from the Spring Brook Intake and Nesbitt Reservoirs and thereby to prevent *Giardia* cysts from entering the finished water supply of residents in the Spring Brook–Hillside Service Area.

Despite these precautions, there was general concern that *Giardia* cysts could find their way to other reservoirs operated by PG&W; indeed, the cysts were found in Elmhurst Reservoir, the principal water supply for Scranton, Pa. This concern prompted some officials to consider whether all water distributed to PG&W customers should be filtered. Because of the fragmented nature of the water supply and distribution systems operated by PG&W, filtration would be costly, both in absolute terms and for individual PG&W customers. One question raised by such a requirement concerns economic efficiency. Do the benefits exceed the costs? Another question concerns equity. Are all PG&W customers able to pay the increased costs of filtered water? This case study addresses only the efficiency issue. A comparison of the benefits and costs of filtration helps place such a requirement in perspective. Although the Spring Brook–Hillside Service Area represents only a small portion of PG&W's extensive water supply system, an analysis of the costs and benefits of constructing a filtration plant near the Spring Brook Intake Reservoir is instructive.

To compare the benefits of preventing future outbreaks of giardiasis in the Spring Brook–Hillside Service Area with the costs of constructing and operating a filtration plant near Spring Brook Intake Reservoir, one needs an estimate of the benefits of preventing an outbreak; the annual probability of occurrence of future outbreaks, assuming that a filtration plant is not constructed (or if constructed, not operated properly); and the costs of constructing and operating a filtration plant. Our analysis assumes that if a waterborne outbreak of giardiasis occurs, *Giardia* cysts are found only in the Spring Brook Intake Reservoir or Nesbitt Reservoir, or both, and not in other water supplies, that the water distribution system for the Spring Brook–Hillside Service Area is the same as it was before the 1983 outbreak, and that the numbers of people and businesses are the same as in 1983. It also assumes that filtration provides complete protection from future outbreaks. Finally, the analysis assumes that no additional benefits, such as less turbid water and a reduced risk of contracting other waterborne diseases, derive from constructing a filtration plant.

The benefits of preventing a future outbreak are defined as the losses avoided. The expected benefits are defined as the average annual benefits (computed as the benefits of preventing an outbreak times the annual probability of an outbreak). The expected benefits and costs of the filtration plant are reported on an annualized basis in 1984 dollars.

One final caveat: the losses from a future outbreak may not be as great as those reported in table 8-1 for the 1983 outbreak. First, a waterborne outbreak of giardiasis would be suspected when the first symptoms of the disease were reported. Local health authorities, the state health and environmental agencies, and PG&W would probably all move faster than they did in 1983 to issue boil-water advisories and to take other protective measures. Second, households would probably arrive at the optimal (least-cost) strategy for obtaining a substitute supply of uncontaminated drinking water sooner than in 1983 because of the experience they gained during the 1983 outbreak. Third, PG&W would be able to determine the source of contamination more quickly and to provide an alternative source of uncontaminated water sooner than it did in 1983, due both to the experience gained during the 1983 outbreak and to the additional flexibility in the water distribution system provided by the additional valves and water mains that were installed then. Fourth, the RFF survey suggests that there might have been some permanent changes in the drinking water habits of some households that would make them less vulnerable to a second outbreak. Finally, evidence in the literature and from physicians in other outbreak areas suggests that some people build up an immunity to giardiasis, so that a second bout with the disease is not as severe as the first; indeed, some people may not experience any symptoms at all.

For all these reasons, a future outbreak of giardiasis in the Spring Brook–Hillside Service Area might not be as severe as the one in 1983. Nonetheless, for purposes of the illustrative benefit–cost analysis in this section and for lack of any better information, the estimates of losses provided in table 8-1 are used to estimate the expected benefits of constructing a filtration plant near Spring Brook Intake Reservoir.

Expected Benefits of a Filtration Plant

The annual expected benefit of preventing future outbreaks of giardiasis in the Spring Brook–Hillside Service Area is the annual probability of a future outbreak (in the absence of a filtration plant) multiplied by the benefits of avoiding an outbreak, or

$$E[B] = pB \qquad (8\text{-}1)$$

where $E[B]$ represents the annual expected benefits, p is the annual probability of a future outbreak, and B denotes the benefits of avoiding an outbreak (table 8-1).

Costs of a Filtration Plant

In 1984, PG&W estimated the cost of constructing a filtration plant to treat water entering the distribution system from the Spring Brook Intake and Nesbitt Reservoirs at $13 million.[2] Using that estimate as a base, the annualized cost of construction at an annual real interest rate of 3 percent is $874,000 for a plant life of twenty years and $663,000 for a plant life of thirty years. At a 5 percent annual real interest rate, the costs are $1,043,000 and $846,000, respectively.

According to PG&W, a minimum of five people would be needed to operate the filtration plant. Together with the costs of chemicals, energy, repairs, and routine maintenance, the annual operating and maintenance costs will amount to at least $500,000.[3] The total annualized costs of the filtration plant at the 3 percent real interest rate would thus be $1,374,000 for a plant life of twenty years and $1,163,000 for a plant life of thirty years. At the 5 percent rate, the costs are $1,543,000 and $1,346,000 respectively.[4]

Benefit–Cost Analysis

None of the three principal pieces of information needed for the benefit–cost analysis is known with certainty. Ranges of estimates are available for the benefits of avoiding a future outbreak and the costs of filtration; no information at all is available on the probability of a future outbreak of giardiasis in the absence of a filtration plant to treat water entering the distribution system from the Spring Brook Intake and Nesbitt Reservoirs.

[2]Pennsylvania Gas and Water Company: Form 10-Q prepared for the Securities and Exchange Commission for the quarter ended March 31, 1984, by Pennsylvania Enterprises, Inc., Wilkes-Barres, Pa., May 1984.

[3]This is a rough estimate based on the costs of salaries and benefits, chemicals, energy, supplies, repairs, and routine maintenance. The estimated annual operating and maintenance costs of a filtration plant at Spring Brook Intake Reservoir were not available from PG&W at the time of writing.

[4]The Nesbitt Water Treatment Plant, which provides for treatment of water from the Nesbitt and Spring Brook Intake Reservoirs, was placed in service in May 1988 (Pennsylvania Gas and Water Company Form 10-K, prepared for the Securities and Exchange Commission for the fiscal year ended December 31, 1989, by the Pennsylvania Gas and Water Company, Wilkes-Barre, Pa., March 1990). The final construction cost for this plant totaled $11.6 million; the operating and maintenance costs (exclusive of depreciation and property taxes) for the year ending August 31, 1990, amounted to $540,708 (personal communication from John F. Kell, Jr., Vice President and Controller, Pennsylvania Gas and Water Company, Wilkes-Barre, Pa., October 16, 1990). The actual total annualized costs of water treatment at this plant, converted to 1984 dollars, are approximately 12 percent less than the estimated total annualized costs used in the illustrative benefit–cost analysis in this chapter. The cost differences between the estimated and actual costs have little impact on the results of the analyses and thus on the conclusions drawn from them. Perhaps more to the point, the benefit–cost analysis in this chapter is an ex ante planning analysis which is made to provide guidance on investments prior to construction. In this context, estimates of benefits and costs are the only information that is ever available.

Given this uncertainty, an unambiguous benefit–cost ratio for a new filtration plant is impossible to compute. However, a benefit–cost analysis based on the range of likely costs, benefits, and probabilities still can be informative.

One way to compare costs, benefits, and the probability of a future outbreak when none can be estimated with certainty is to hold one of the variables constant and to explore the effect on the benefit–cost ratio of variations in the other two. Another approach is to hold either the benefits or the costs constant and compute the minimum annual probabilities of an outbreak for the expected benefits to be equal to or greater than the costs. A third approach is to vary all three variables simultaneously and to identify those combinations of benefits, costs, and probabilities of a future outbreak where the expected benefits are equal to or greater than the costs. The first two approaches lend themselves to tabular presentation of the results. The third approach is best presented graphically.

An example of each of these benefit–cost comparisons is provided to illustrate different ways of communicating the results of benefit–cost analyses for those who plan to use them in decisions. One form of presentation may be more useful than another, depending on the audience and on the variable of most concern for a particular decision. First, an annual probability of an outbreak in the absence of a filtration plant is assumed, and the annual expected benefits and the annualized costs of filtration are compared (using benefit–cost ratios) for the six loss estimates provided in table 8-1 and the four cost estimates presented above. Second, the six estimates of the benefits of avoiding an outbreak are compared with the largest (most conservative) estimate of the annualized costs of filtration, and then the annual probabilities (and average return periods) of an outbreak needed to make the annual expected benefits equal to or greater than the annualized costs are computed. (The average return period in years, a term often used to describe the frequency of floods and other natural hazards, is the reciprocal of the annual probability of an outbreak.) Third, a graphical representation of the relationship among the benefits of avoiding an outbreak, the annual probability of a future outbreak, and the annualized costs of filtration is provided, which shows clearly those combinations of benefits, costs, and probabilities where the annual expected benefits of filtration are equal to or greater than the annualized costs.

First Analysis The first analysis assumes that there is a 10 percent annual probability (average return period of ten years) of a future outbreak if the filtration plant is not constructed. (Based on the information available, a 10 percent annual probability of a future outbreak is no more or less likely than either a 2 percent or a 20 percent annual probability of a future outbreak.) For this analysis, the annual expected benefits, $E[B]$, from expression (8-1), are computed for the six estimates of benefits of avoiding an outbreak, B, reported as losses in table 8-1, and for $p = 0.10$. Then the annual expected benefits are compared with the four estimates of the annualized costs of filtration reported above. Benefit–cost ratios for twenty-four combinations of benefits and costs are presented in table 8-5.

Based on an assumed 10 percent annual probability of a future outbreak if the filtration plant is not built, and on the firm lower-bound estimate of the benefits of avoiding an outbreak—$9.2 million (scenario C, group I), construction of a filtration plant near Spring Brook Intake Reservoir would not be justified on the basis of a benefit–cost analysis. The benefit–cost ratios for four estimates of the costs of the filtration plant range between 0.6 and 0.8 (table 8-5). (Recall that a benefit–cost analysis addresses only the economic efficiency issue, which is just one consideration in a decision to build a filtration plant.) Nor would construction of a filtration plant be justified using the next highest estimate of benefits—$10.0 million (scenario B, group I). The benefit–cost ratios for four estimates of the costs of the filtration plant range between 0.6 and 0.9 (table 8-5). The benefits of avoiding an outbreak in this illustrative example would have to be greater than $15.4 million ($1.54 million/0.10 = $15.4 million) for all four benefit–cost ratios to be greater than 1.0.

Table 8-5. Comparison of the Benefits and Costs of a Filtration Plant near Spring Brook Intake Reservoir for 10 Percent Annual Probability of a Future Outbreak

| | | | Benefit–cost ratio | | | |
| | Benefits of avoiding an outbreak[b] (millions of | Annual expected benefits[c] (millions of | Range of annualized costs of filtration (millions of 1984 dollars) | | | |
Benefit scenario[a]	1984 dollars)	1984 dollars)	$1.16	$1.35	$1.37	$1.54
Group I						
C	9.24	0.924	0.8	0.7	0.7	0.6
B	9.97	0.997	0.9	0.7	0.7	0.6
A	25.76	2.576	2.2	1.9	1.9	1.7
Groups I and II						
C	23.27	2.327	2.0	1.7	1.7	1.5
B	27.27	2.727	2.4	2.0	2.0	1.8
A	55.47	5.547	4.8	4.1	4.0	3.6

[a]From table 8-1 and described in table 8-2.
[b]From table 8-1.
[c]Based on equation (8-1) and on an assumed 10 percent annual probability of a future outbreak in the Spring Brook–Hillside Service Area in the absence of a new filtration plant to treat water entering the distribution system from Spring Brook Intake and Nesbitt Reservoirs.

Four estimates of benefits in table 8-5 produce annual expected benefits greater than the annualized costs of a filtration plant. The two estimates closest to $15.4 million are

- $23.3 million (scenario C, groups I and II), and
- $25.8 million (scenario A, group I).

To increase total benefits above the firm lower-bound estimate of $9.2 million requires additional benefit categories or additional assumptions, or both. For the $23.3 million estimate, a number of categories of benefits must be added to the firm lower-bound estimate. These additional categories, shown in table 8-1, include the value of:

- lost productivity and lost leisure time activities due to illness;
- reductions in tap water for drinking, food preparation, and personal hygiene beyond the quantity supplied by alternative sources of drinking water;
- lost profits and lost consumer surplus associated with restaurants and bars in the outbreak area; and
- time premiums and possible costs associated with schools, other businesses, and government agencies.

To reach a benefit estimate of at least $15.4 million, only the second category, or a combination of the first and second categories, of benefits would be required.

For the $25.8 million estimate of benefits, categories of benefits do not have to be added to the firm lower-bound estimate, but different assumptions concerning the value of time must be made. These assumptions, summarized in table 8 2 for scenario A (group I), include the value of:

- time to obtain medical treatment, for homemakers and retirees (increased from $0 per hour to $6.39 per hour);
- lost work, for homemakers and retirees (increased from $0 per hour to $6.39 per hour, 40 hours per week);
- time to haul water to households, for homemakers and retirees (increased from $0 per hour to $6.39 per hour);
- time to purchase water for household use, for those employed, homemakers, and retirees (incorporated in the assumed price of bottled water, increased from $0.66 per gallon to $1.17 per gallon);
- time to boil water for household use, for those who are employed, homemakers, and retirees at $6.39 per hour, and teenagers at $2.65 an hour (changed from a combined activity to a sole activity);
- time to haul water to restaurants and bars (increased from $3.35 per hour to $5.00 per hour); and
- time to boil water at restaurants and bars at $5.00 per hour (changed from a combined activity to a sole activity).

To reach a benefit estimate of at least $15.4 million, not all these assumptions are necessary.

The assumptions made in estimating losses and the number of loss categories included in the totals greatly influence estimates of total benefits, benefit–cost ratios, and guidance for policy. As shown in table 8-5, of the twenty-four benefit–cost ratios, eight are less than 1.0 and sixteen are greater than 1.0. Overall, the benefit–cost ratios in table 8-5 range from 0.6 to 4.8, representing an eightfold difference between the lowest and the highest estimate.

Wide variations in estimates of benefits and costs can be a problem for policy guidance. But the principal problem with this analysis is that it depends critically on the assumption of the annual probability of a future outbreak in the absence of a filtration plant. The issue for policy in the absence of any better data is the following: Is an average return period of ten years for an outbreak of giardiasis reasonable, or is it likely to be shorter or longer than that? If the average return period is shorter than ten years, a filtration plant may be indicated on the basis of a benefit–cost analysis. If it is longer than ten years, it may not be. This issue requires further analysis.

Second Analysis   This analysis computes what the annual probability (and average return period) of a future outbreak would have to be for the annual expected benefits of the filtration plant to be equal to or greater than the annualized costs of constructing and operating it. The benefits of avoiding an outbreak, B, reported as losses in table 8-1, and a conservative (high) estimate of the annual costs of filtration—$1.54 million per year (twenty-year plant life, 5 percent real interest rate) are used.

For this analysis, the annual expected benefits, $E[B]$, are set equal to or greater than the annualized costs, C.

$$E[B] \geq C \qquad (8\text{-}2)$$

Using expressions (8-1) and (8-2), the benefits of avoiding an outbreak, B, can be expressed in terms of the annualized costs, C, and the annual probability of a future outbreak, p.

$$pB \geq C \qquad (8\text{-}3)$$

The annual probability of an outbreak for the expected benefits to be equal to or greater than the costs may be computed from the benefits of avoiding an outbreak, B, and the annualized costs of filtration, C, according to the following expression derived from (8-3).

$$p \geq \frac{C}{B} \tag{8-4}$$

The minimum annual probabilities and the maximum average return periods (in years) that ensure the annual expected benefits of a filtration plant are equal to or greater than the annualized costs, for the losses of an outbreak reported in table 8-1, are shown in table 8-6.

The maximum average return periods for an outbreak of giardiasis range from 6.0 years to 16.7 years for the group I estimates of benefits, and from 15.1 years to 36.0 years for the sum of the group I and group II estimates of benefits. If the actual average return period for an outbreak is less than that shown in table 8-6, the expected benefits would *exceed* the costs, and thus the construction of the filtration plant would be indicated on the basis of a benefit–cost analysis.

Third Analysis The relationship among the benefits of avoiding a future outbreak, the costs of the filtration plant, and the annual probabilities of a future outbreak is depicted graphically in figure 8-1. The three linear relationships (diagonal lines) shown in this figure represent

Table 8-6. Minimum Probabilities and Maximum Return Periods of a Future Outbreak Required for the Expected Benefits of a Filtration Plant near Spring Brook Intake Reservoir to be Equal to or Greater Than the Costs

Benefit scenario[a]	Benefits of avoiding an outbreak[b] (millions of 1984 dollars)	Minimum annual probability[c]	Maximum average return period[d] (years)
Group I			
Scenario C	9.24	0.167	6.0
Scenario B	9.97	0.154	6.5
Scenario A	25.76	0.060	16.7
Groups I and II			
Scenario C	23.27	0.066	15.1
Scenario B	27.27	0.056	17.7
Scenario A	55.47	0.028	36.0

[a]From table 8-1 and described in table 8-2.
[b]From table 8-1.
[c]Based on expression (8-4) for the annualized costs of filtration, C, equal to $1.54 million per year.
[d]The average return period in years is the reciprocal of the annual probability, p.

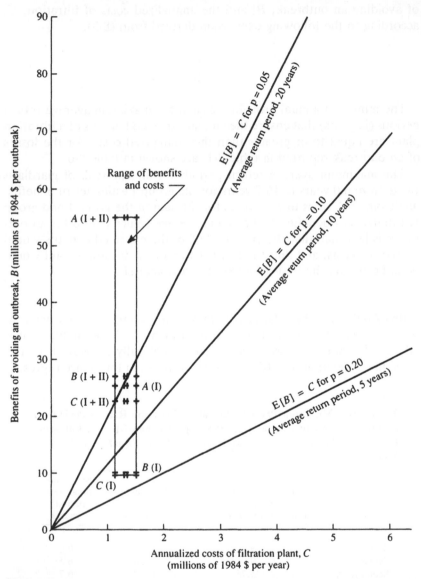

Figure 8-1. Comparison of the benefits of avoiding an outbreak and the annualized costs of filtration for different annual probabilities of a future outbreak.

conditions where the annual expected benefits equal the annualized costs of filtration for three annual probabilities of a future outbreak—20 percent (average return period of five years), 10 percent (average return period of ten years), and 5 percent (average return period of twenty years). Points above each line represent conditions where the annual expected benefits of a filtration plant are greater than the annualized costs for the given probability. Points below the line represent conditions where the annual expected benefits are less than the annualized costs.

For an average return period of twenty years, the sum of the group I and II estimates of annual expected benefits for scenario A exceeds the annualized costs; for scenario B it is about equal to the annualized costs; and for scenario C it is less than the annualized costs (with one exception involving the lowest cost estimate). The group I estimate of annual expected benefits for scenario A is about equal to the annualized costs. The group I estimates of annual expected benefits for scenarios B and C are less than the annualized costs.

For an average return period of ten years, the group I estimate of annual expected benefits for scenario A is greater than the annualized costs, and for scenarios B and C the annual expected benefits are less than the annualized costs (see also table 8-5). For an average return period of five years, all three group I estimates of annual expected benefits are greater than the annualized costs. Thus, if the average return period for an outbreak of giardiasis is as short as five years, there is little doubt that the expected benefits of a filtration plant would exceed the costs of constructing and operating it. If the average return period is as long as twenty years, substantial confidence would need to be placed in the group II estimate of benefits for scenario B and in the lower estimate of the costs of filtration to ensure that the annual expected benefits of filtration exceed the annualized costs.

Because no data are available on the annual probability of a future outbreak of giardiasis in the Spring Brook–Hillside Service Area in the absence of a filtration plant to treat water entering the distribution system from the Spring Brook Intake and Nesbitt Reservoirs, a more precise benefit–cost analysis cannot be made. Nonetheless, the results of the three analyses, particularly the graphical representation in figure 8-1, do place the benefits and costs of constructing a new filtration plant in perspective.

LESSONS FROM THE CASE STUDY

The estimates of the benefits of avoiding the 1983 outbreak of giardiasis in Luzerne County and the comparison of these benefits with the costs of filtration demonstrate that benefit analysis and benefit–cost analysis can provide useful information upon which to base decisions on protecting the public from contaminated drinking water. On the basis of the three benefit–cost analyses presented in the last section, a filtration plant to treat water entering the distribution system from the Spring Brook Intake and Nesbitt Reservoirs cannot be supported unless the average return period of a future outbreak in the absence of a filtration plant is as short as five years, or, assuming it is as long as ten to fifteen years, unless the estimates of benefits represented by scenario A in group I or scenario C in groups I and II can be justified.

This assessment is based on conservative estimates of the benefits of avoiding future outbreaks; not all the losses of the 1983 outbreak were included in the estimates of benefits. Had losses such as pain and suffering, and benefits such as reduced levels of turbidity been included in the estimates of benefits, the benefit–cost analysis in the last section would have been more favorable toward constructing a filtration plant.

Finally, it cannot be assumed from a single case study that all water supply systems are similar to the Spring Brook–Hillside system, and that all public water supplies, therefore, should or should not be provided with filtration. There are differences in the risk of contamination of the many watersheds in the United States (indeed, even of those within the water supply system operated by PG&W), differences in the populations exposed to the various water supplies, differences in the kinds of water supply system failures associated with other outbreaks of giardiasis, and differences in the per capita costs of water treatment, all of which have an impact on the benefits and costs of filtration in particular situations. For some water supply systems, the benefits will exceed the costs. For other water supply systems, particularly the smaller systems with low risks of contamination, they will not. More information on the factors placing a water supply at risk for *Giardia*, on the benefits of avoiding waterborne giardiasis in particular situations, and on the per capita costs of treatment for different-sized systems is needed to determine if investment in filtration is efficient for a particular water supply system.

This study provided information that the Environmental Protection Agency needed to assess the economic efficiency of proposed drinking water regulations (see chapter 9). It developed methods and survey instruments for estimating the economic losses associated with an outbreak of waterborne giardiasis and tested those methods and instruments in the field by applying them to a case study of a known outbreak. These methods and instruments can be used to estimate the losses associated with other known outbreaks of giardiasis as well as the benefits of preventing future outbreaks. They also can be used to estimate the benefits of avoiding other episodic waterborne diseases where morbidity or averting behavior, or both, play significant roles. Together with information on the costs of water treatment and on the probabilities of future outbreaks of waterborne giardiasis in the absence of treatment, these methods permit an assessment of the benefits and costs of proposed drinking water regulations.

Part 3 / Research Needs

9 / Research Needs

This study has made progress on several fronts in estimating the benefits of avoiding an outbreak of a nonfatal waterborne disease—in this case, giardiasis. First, it developed theoretical models (1) to estimate losses to individuals who contracted giardiasis and thereby incurred expenses for medical treatment, lost time at work, and suffered diminished enjoyment of leisure activities; and (2) to estimate losses associated with activities taken to avoid drinking contaminated water. Second, the study developed measurement instruments used to collect data on the effects of an outbreak on individuals, businesses, and government agencies. Third, it used the theory and the measurement instruments to estimate the losses caused by an actual outbreak of giardiasis.

This study also has made progress on estimating the benefits of preventing a specific outbreak of giardiasis. But the extent to which it can be used to estimate the benefits of preventing such outbreaks in other areas is more problematic because it is not known whether the Luzerne County outbreak is typical of other actual and potential outbreaks. Furthermore, the study, by itself, cannot be used to determine whether and where investments in water treatment technology are warranted. Information on treatment costs is needed for this, as well as information on the probability that an outbreak would occur in the absence of water treatment. For making decisions on whether new regulations for the protection of public health are indicated, similar, but more broadly based, information is needed, as well as information on the frequency of occurrence of giardiasis in the absence of water treatment.

In fact, much of this information was collected and used, along with this benefit study, by the Environmental Protection Agency to analyze alternative regulatory strategies for protecting the public from *Giardia*-contaminated municipal water and to devise and help justify a recent regulation (54 Fed. Reg. 27486 [June 29, 1989]).

The usefulness of benefit analysis for the policy applications described above depends in part on the accuracy of the estimates. Inasmuch as the estimates of total losses from the Luzerne County outbreak range between a firm lower bound of $9 million and a high estimate of $55 million (in 1984 dollars)—approximately a sixfold difference—improvements in the accuracy of such estimates would seem to be indicated. The need for improved estimates in particular cases, however, depends on the estimates of the costs of water treatment and on other factors, as demonstrated in chapter 8.

The following areas contributed most to the uncertainty in the estimates of losses for Luzerne County and, hence, are important candidates for future research:

- the design of the questionnaires,
- the extrapolation of survey results to the affected population,
- the valuation of time,
- the valuation of illness, and
- the valuation of anxiety.

QUESTIONNAIRE DESIGN

Although we were generally pleased with the results of the three main data-gathering instruments (questionnaires for the survey of individuals with confirmed cases of giardiasis, for the household survey on averting behavior, and for the survey of restaurants and bars, contained in appendixes 6-A, 6-B, and 7-A, respectively, in this volume), there is much room for improvement. The questionnaire for individuals with confirmed cases of giardiasis, for example, did not elicit useful information on "special" leisure activities; that is, the highly valued leisure activities that cannot be rescheduled if illness occurs. Most respondents mentioned one or two special activities (in one case, a honeymoon) that were cancelled or ruined by the illness. However, we felt it too burdensome to ask respondents to answer all the questions we would like to have asked.

The results of the "productivity loss" questions might also be suspect. Respondents may have been describing how they felt physically rather than assessing their loss of productivity. Because so many victims of giardiasis continued their normal activities for most of the duration of their illness, the productivity issue has important economic implications, and the questionnaire needs to be improved to obtain more useful information in this area.

In the household survey on averting behavior, time constraints were imposed on the interviews to avoid taxing the patience of the interviewees. As a result, averting behaviors may have been oversimplified. In addition, data needed to estimate the consumption of water by specific use before the outbreak were unavailable; national per capita consumption data were used instead. Finally, little information was gathered about the substitution of other liquids for drinking water after the public water supply was implicated.

The restaurant and bar questionnaire elicited the information one could reasonably expect from businesses. However, the 9 percent return rate of this questionnaire was disappointingly low, especially in view of the 49 percent response rate for the confirmed-case questionnaire. Techniques for raising the business response rate in an affected area should be investigated.

EXTRAPOLATION OF SURVEY RESULTS

Problems were encountered in extrapolating the results of the three surveys to the total population in the affected area. The most troublesome was the use of average losses due to illness incurred by the confirmed cases, minus hospitalization costs, to estimate the losses incurred by the clinical cases. The principal concern is that the average loss of the clinical cases might have been less than that of the confirmed cases. This procedure could be improved by asking a sample of persons with clinical symptoms but without confirmed diagnoses to complete the same questionnaire sent to the people with confirmed cases. This procedure would uncover any significant differences between the two groups in losses due to illness.

The principal difficulty encountered with the restaurant and bar survey was the small sample size, which, together with the wide variation of responses, made extrapolation to all restaurants and bars in the outbreak area subject to considerable error.

VALUATION OF TIME

A major portion of the total losses arose from the imputed value of time, rather than from out-of-pocket expenses. Thus, the valuation of time is critical. Time losses figured prominently in estimates of the value of lost work and leisure time associated with illness, as well as losses associated with obtaining alternative water supplies. For lack of any better information, lost time was valued at the individual's wage rate, in some cases adjusted for taxes. For lost work time, such a valuation requires the tenuous assumption that the wage rate equals the marginal value of an individual's labor. Using the after-tax wage rate to value leisure-time activities assumes that workers are free to choose the quantity of work, even though the length of the work day is fixed for most people. Moreover, even with this assumption, the wage rate applies only at the margin (the last hour of work and the last hour of leisure activity). It is not clear how to value the inframarginal leisure hours likely to be at stake, especially because the value of the first leisure hour available is likely to be considerably higher than the wage rate. Finally, not everyone is a wage-earner. The value of time of homemakers, retirees, the unemployed, and even children is certain to be greater than zero, yet only for homemakers is there even a remote market proxy.

Making progress on the valuation of time will be difficult, but the potential payoff in improved estimates of benefits would be great indeed.

VALUATION OF ILLNESS

The effects of illness extend beyond the time taken away from other activities. Illness also imposes direct discomfort that individuals would presumably pay to avoid. While indirect measures of this "pain and suffering" component may eventually emerge, useful information on discomfort is more likely to be obtained from direct questioning, as in a contingent valuation (CV) study. CV research on health impacts, which is for the most part concerned with the willingness to pay to reduce the risks of particular diseases, or the willingness to pay to take some policy action to reduce illness incidence, may therefore have to be redirected or expanded. More useful for morbidity, but not mortality, studies would be the willingness to pay to avoid the actual consequences of a particular disease (rather than the increased probability of contracting it).

VALUATION OF ANXIETY

The survey of individual averting behavior pointed to one final area where research is indicated. More than half of those who responded to the survey reported a permanent change in drinking water habits, even though the water supply had been declared safe by the Pennsylvania Department of Environmental Resources. The lack of faith in the safety of the local water supply is itself a costly outcome of the Luzerne County outbreak, with additional losses extending far beyond the period of the outbreak. It would be interesting to know whether such behavior was solely the result of the outbreak or whether, as some respondents suggested, the outbreak was "the last straw" in a series of incidents involving the local water supply. In either case, it would be interesting to determine if completion of the filtration plant near the Spring Brook Intake Reservoir restored trust in the local water supply.

ABOUT THE AUTHORS

Winston Harrington is a senior fellow in the Quality of the Environment Division at Resources for the Future, and currently a visiting professor in economics at the University of California at Santa Barbara. He received his Ph.D. in city and regional planning from the University of North Carolina at Chapel Hill in 1985. He is a coauthor of the RFF book *Enforcing Pollution Control Laws*, and has written numerous articles on a wide range of issues in environmental economics and policy.

Alan J. Krupnick is a senior fellow in the Quality of the Environment Division at Resources for the Future and head of RFF's Energy and Environment Program. He received his Ph.D. in economics from the University of Maryland in 1980. He is coauthor of the RFF book *Rules in the Making: A Statistical Analysis of Regulatory Agency Behavior*, and has written numerous papers on the valuation of health and environmental improvements and on benefit–cost analyses of projects and policies to improve the environment.

Walter O. Spofford, Jr., is a senior fellow in the Quality of the Environment Division at Resources for the Future, and a former director of that division. Currently he is director of RFF's Environment and Development Program. He received his Ph.D. in water resources engineering from Harvard University in 1965. An environmental systems engineer-economist with more than twenty years' experience in regional environmental economic analysis, Spofford has written numerous papers on the economics of regional environmental quality management. He also has extensive international consulting experience. He has assisted various countries, including the People's Republic of China, with water resources management and development, regional environmental quality management, and integrating environmental considerations in development planning.

Index

Library of Congress Cataloging-in-Publication Data

Harrington, Winston.
    Economics and episodic disease : the benefits of preventing a
giardiasis outbreak / Winston Harrington, Alan J. Krupnick, and
Walter O. Spofford, Jr.
      p.   cm.
    "The product of RFF's Quality of the Environment Division"—T.p.
verso.
    Includes bibliographical references and index.
    ISBN 0-915707-59-4 (alk. paper)
    1. Giardiasis—Economic aspects—Pennsylvania—Luzerne County.
2. Giardiasis—Epidemiology—Case studies. 3. Waterborne infection—
Prevention—Economic aspects—Case studies.   I. Krupnick, Alan J.
II. Spofford, Walter O.   III. Resources for the Future. Quality of
the Environment Division. IV. Title.
RA644.G53H37   1991   91-2129
363.6'1—dc20   CIP